ELEMENTS OF THE
THEORY OF FUNCTIONS

BY KONRAD KNOPP

Professor of Mathematics at the University of Tübingen

TRANSLATED BY FREDERICK BAGEMIHL

DOVER PUBLICATIONS, INC.

Published in Canada by General Publishing Company, Ltd., 30 Lesmill Road, Don Mills, Toronto, Ontario.

Published in the United Kingdom by Constable and Company, Ltd., 10 Orange Street, London W. C. 2.

This Dover edition, first published in 1952, is a new English translation of *Elemente der Funktionentheorie*.

International Standard Book Number: 0-486-60154-4
Library of Congress Catalog Card Number: 53-7031

Manufactured in the United States of America
Dover Publications, Inc.
180 Varick Street
New York, N. Y. 10014

CONTENTS

Chapter XIII. The Logarithm, the Cyclometric Functions, and the Binomial Series

ELEMENTS OF THE
THEORY OF FUNCTIONS

CHAPTER I

FOUNDATIONS

1. Introduction

The name *"Theory of Functions"* is used to denote all those investigations which arise if one seeks to transfer the problems and methods of real analysis (i.e., differential and integral calculus, and related fields) to the case in which all numerical quantities (constants, independent and dependent variables) that appear are permitted to be complex numbers, that is, numbers of the form $a + b\sqrt{-1}$. Early and quite automatically such considerations forced their way into investigations connected with various problems in real analysis, and have been carried out, in conjunction with the solution of these problems, in the course of centuries,—hesitantly at first, but soon with ever greater success (see §4 for further details). Today the theory of functions is one of the most extensive and important branches of higher mathematics.

In these *Elements of the Theory of Functions* we shall treat only those topics which are simplest, but which are at the same time most important for the further development of the theory.[1] This includes, first, an introduction to the system of complex numbers and the operations performed on them. Then, the

[1]This development is given by the present author in the following two little volumes: *Theory of Functions, Part I: Elements of the General Theory of Analytic Functions*, translated from the 5th German edition, New York, 1945; and *Theory of Functions, Part II: Applications and Continuation of the General Theory*, translated from the 4th German edition, New York, 1947. We shall refer to these volumes, in what follows, as *"Th. F. I"* and *"Th. F. II,"* for brevity.

concept of sets of numbers, the limit concept, and closely related matters, in particular the theory of infinite series, are extended to complex quantities, or, as we say briefly, "to the complex domain." Further, we shall carry over the notion of function and its most important properties to the case in which independent and dependent variables are complex. When combined with the limit concept, this yields the foundations of a differential calculus for functions of a complex variable. Finally, we shall study more closely the so-called elementary functions, including the rational, and, in particular, the linear, functions, the exponential function, the trigonometric functions, and several others, as well as their inverses, such as the logarithm and the cyclometric functions. The extension of the integral calculus to the complex domain, however, is not regarded as belonging properly to the elements of the theory of functions.

We shall see (chs. 2 and 3) that the operations performed on complex numbers, and, eventually, that all the investigations just mentioned, can be visualized, in a *number plane* or on a *number sphere*, even more vividly than in the "real domain". This forms the content of that part of our theory which is called *"geometric theory of functions"*.

It is evident from what has been said, that in order to be able to understand this little book, a knowledge of the foundations of real analysis and of the elements of analytic geometry is indispensable, inasmuch as the extension to the complex domain is accomplished after the pattern of real analysis, and use is made of simple geometric facts for purposes of visualization. So as to have a fixed point of departure, we shall state in §2 what is most important concerning the system of real numbers, which constitutes the foundation for the erection of real analysis, and shall discuss in §3 what is most fundamental as regards the construction of analytic geometry.

2. *The system of real numbers*

We presuppose familiarity with the system of real numbers, of course, as far as its practical use is concerned; but because of its fundamental importance, we shall present briefly here the essential ideas which lead to its construction.

The starting point of all investigations concerning numbers is the sequence of *natural numbers*, 1, 2, 3, . . . , and the two "operations" on them: *addition* and *multiplication*. The necessity of performing the "inverses" of these operations presently compels the introduction of 0 (zero), negative numbers, and, finally, fractions. The totality of integral, fractional, positive and negative numbers, and zero, is called the *system of* (real) *rational numbers*.

One can operate with these numbers, which are now denoted for brevity by single Roman letters, according to certain rules which are called the *fundamentals laws of arithmetic*. These are the following, in which by "numbers" we mean, for the present, only the rational numbers just referred to:

I. Fundamental laws of equality and order

1. *The set of numbers is an ordered set; i.e., if a and b are any numbers, they satisfy one, and only one, of the relations*[2]

$$a < b, \qquad a = b, \qquad a > b.$$

This order obeys these additional laws:

2. $a = a$ *for every number a.*
3. $a = b$ *implies* $b = a$.
4. *If* $a = b$ *and* $b = c$, *then* $a = c$.
5. *If* $a \leqq b$ *and* $b < c$, *or if* $a < b$ *and* $b \leqq c$, *then* $a < c$.

All numbers which are greater than zero are called *positive*, all numbers which are less than zero are called *negative*. If a number is equal to zero, we also say that it *"vanishes"*.

II. Fundamental laws of addition

1. *Every pair of numbers a and b can be added; the symbol* $(a + b)$ *or* $a + b$ *always represents a definite number, the sum of a and b.*

---

[2]Read: a is less than b, a is equal to b, a is greater than b. $a > b$ is merely another way of writing the relation $b < a$.

The negations of these three relations are written as follows:

$a \geqq b$ (a is greater than or equal to b, a is at least as great as b, a is not less than b),

$a \neq b$ (a is not equal to b),

$a \leqq b$ (a is less than or equal to b, a is at most as great as b, a is not greater than b).

This formation of sums obeys these laws:

2. *If $a = a'$ and $b = b'$, then $a + b = a' + b'$.* ("If equals are added to equals, the sums are equal.")

3. $a + b = b + a$. (Commutative law.)

4. $(a + b) + c = a + (b + c)$. (Associative law.)

5. $a < b$ *implies* $a + c < b + c$. (Monotonic law.)

III. FUNDAMENTAL LAW OF SUBTRACTION

The inverse of addition can always be performed; i.e., if a and b are any numbers, there exists a number[3] x such that $a + x = b$.

The number x thus determined is called the *difference* of b and a, and is denoted by $(b - a)$.

IV. FUNDAMENTAL LAWS OF MULTIPLICATION

1. *Every pair of numbers a and b can be multiplied; the symbol $a \cdot b$ or ab always represents a definite number, the product of a and b.*

This formation of products obeys these laws:

2. *If $a = a'$ and $b = b'$, then $ab = a'b'$.* ("If equals are multiplied by equals, the products are equal.")

3. $ab = ba$. (Commutative law.)

4. $(ab)c = a(bc)$. (Associative law.)

5. $(a + b)c = ac + bc$. (Distributive law.)

6. *If $a < b$, and $c > 0$, then $ac < bc$.* (Monotonic law.)

The four "rules of sign" and, as a supplement to them, the result that

$$a \cdot 0 = 0 \text{ for every number } a,$$

follow in the simplest fashion, but certainly as demonstrable facts, from the fundamental laws enumerated thus far. The four rules of sign assert, in particular, that

$$\text{if } a \neq 0 \text{ and } b \neq 0, \text{ then } ab \neq 0.$$

From this and the preceding result follows the important

THEOREM. *A product of two numbers is equal to zero if, and only if, at least one of the two factors is equal to zero.*

[3] We need not assume that this number x be *uniquely* determined by a and b, as it follows easily from the remaining fundamental laws; in particular, II, 5.

V. Fundamental law of division

The inverse of multiplication can, except in one case, always be performed; i.e., if a and b are numbers, the first of which is not equal to zero, there exists a number[4] x such that $ax = b$.

The number x thus determined is called the *quotient* of b and a, and is denoted by b/a.

All these laws can be deduced very easily from the most elementary properties of the natural numbers. Now, the importance of listing them is this: Once the validity of these fundamental laws has been established, it is unnecessary, in all further work with the literal quantities $a, b, \ldots$, to make use again of the fact that these symbols denote rational numbers. All further rules of operation can be inferred purely formally,[5] with complete rigor, from the validity of the fundamental laws alone. Such rules have already been mentioned in IV. They include, in addition, all so-called rules of parentheses, the manipulation of equalities and inequalities, in short, all rules of the so-called literal calculus, into which we shall, of course, not enter further here.

From the important fact that the *meaning* of the literal symbols need not be considered at all in this connection, there results immediately the following extraordinarily significant consequence: If one has any other entities whatsoever besides the rational numbers,—we shall mention such other entities presently,—*but which obey the same fundamental laws*, one can operate with them as with the rational numbers, according to exactly the same rules. Every system of objects for which this is true is called a *number system*, because, in a few words, it is customary to call all those objects *numbers* with which one can operate according to the fundamental laws we have listed.

Such other entities which also obey all our fundamental laws are, in particular, the *real numbers*. We recall briefly how one arrives at them. The system of rational numbers is incomplete in the sense that it is incapable of satisfying very simple demands. Thus, as is well known, there is no rational number

[4]Here, as in the case of subtraction, x is uniquely determined by a and b.
[5]I.e., without having to consider the *meaning* of the symbols.

whose square is equal to 2. The fact that a rational number exists whose square is as close to 2 (greater or less than) as one pleases, together with the familiar representation of the state of affairs on a number axis (see §3 for further details), leads one to divide all rational numbers into two classes: a class $\mathfrak{A}$, which contains zero, the negative rational numbers, and every positive rational number whose square is less than 2; and a class $\mathfrak{A}'$, which contains every positive rational number whose square is greater than 2. The "irrational" number whose square is equal to 2 is said to be realized by means of this classification, or this *Dedekind cut*, $(\mathfrak{A}|\mathfrak{A}')$, in the domain of rational numbers, and one actually writes $(\mathfrak{A}|\mathfrak{A}') = \sqrt{2}$.

That such a Dedekind cut really defines a number or even *is* a number can be proved only in the following manner: One considers the *totality of all conceivable divisions* of the rational numbers into two (non-empty) classes $\mathfrak{A}$ and $\mathfrak{A}'$ which, as above, satisfy the requirement that *every number of the class $\mathfrak{A}$ be less than every number of the class $\mathfrak{A}'$*. Then one shows that these *Dedekind cuts* $(\mathfrak{A}|\mathfrak{A}')$ are such "other entities" which, when suitable stipulations are made as to the meanings of the symbols $=$, $<$, $+$, and $\cdot$, again satisfy all our fundamental laws. How these stipulations are to be chosen and how the proof in question can be furnished will not be considered here, but will be regarded as familiar to the reader.[6] The way to proceed is obvious when the matter is viewed on the number axis. If one now denotes these cuts for brevity by small Roman letters, sets $(\mathfrak{A}|\mathfrak{A}') = a$, etc., and calls them numbers, then under these agreements all our fundamental laws hold without exception. The entities obtained in this manner therefore *are* numbers. They, in their totality, constitute the *system of real numbers*. When they are represented on the number axis, it turns out that some of the real numbers coincide with the hitherto existing rational numbers, and some do not. In this sense the system of real numbers is an *extension* of the system of rational numbers. Those real numbers which are not rational are called *irrational*.

[6]Consult the works of the present author, or of Perron, Hardy (Ch. I), or Bromwich (App. I), listed on p. 136.

With the construction of the system of real numbers, a certain closure is now attained. For, it is possible to show that no different system (distinct, in any essential respect, from the acquired system of real numbers), and no more extensive system, of any entities whatsoever, exists, which satisfies *all* our fundamental laws—no matter how the meanings of the symbols $=$, $<$, $+$, and   be defined. The theorems indicated herewith are known as the *uniqueness theorem* and the *completeness theorem*, respectively, for the system of real numbers.

A renewed classification in the system of real numbers, instead of giving rise to new entities, always leads to an already existing real number. Thus, if one makes another Dedekind cut in the domain of real numbers, i.e., if one divides *all real numbers* into two (non-empty) classes $\mathfrak{A}$ and $\mathfrak{A}'$ such that every number a of $\mathfrak{A}$ is less than every number a' of $\mathfrak{A}'$, then the following *theorem of continuity* for the real numbers, often called the *fundamental theorem of Dedekind*, can be proved:

THEOREM. *Such a Dedekind cut in the domain of real numbers always defines one, and only one, real number, s, the "cut-number," such that every $a \leqq s$, every $a' \geqq s$.*

The cut-number, s, itself may belong to $\mathfrak{A}$ or to $\mathfrak{A}'$, depending on the classificatory viewpoint. Every number less than s, however, belongs to $\mathfrak{A}$, every number greater than s, to $\mathfrak{A}'$.

3. Points and vectors of the plane

In what follows, we require only the simplest and most familiar of the fundamental concepts of analytic geometry, and we therefore confine ourselves to the presentation of those principles which are of greatest importance for its erection.

First, the *rational* numbers may be represented in the well-known manner by points of a *number axis*, i.e., an arbitrary straight line (imagined horizontal) on which two distinct points O and U have been chosen as origin and unit point, or, briefly, as 0 and 1, respectively, 1 to the right of 0. In this way, the fact that the rational numbers form an ordered set becomes graphically clear.

The considerations contained in §2 now show that to every rational number there corresponds precisely one point—we call it, for brevity, a rational point,—but that not every point of the straight line is the image of a rational number. To every Dedekind cut in the domain of rational numbers, however, there corresponds a cut in the set of rational points: the latter are divided into two (non-empty) classes $\mathfrak{A}$ and $\mathfrak{A}'$ such that every point a of $\mathfrak{A}$ lies to the left of every point a' of $\mathfrak{A}'$. Intuition is imperative here, and demands that there always exist a point s on the line, which separates the two classes, i.e., which is such that $a \leq s \leq a'$ for all a and a'. The explicit recognition of this fact forms the content of the

CANTOR-DEDEKIND AXIOM. *Every cut in the domain of rational points defines a unique point of the straight line, which separates the two classes of the cut.*

This means merely that to every real number there corresponds precisely one point of the straight line as its image, and conversely. In this sense the system of real numbers is in one-to-one correspondence with the points of the number axis. Considering this correspondence, the *theorem of continuity* stated at the end of §2 says the following:

If all points of the number axis are divided in any manner into two non-empty classes such that every point of the first class lies to the left of every point of the second class, then there is always precisely one point which separates the two classes.

This correspondence between the real numbers and the points of a straight line is the foundation of analytic geometry. Instead of representing the real numbers by means of the points of the number axis, it is sometimes more advantageous to represent them by means of the directed segments, the *vectors*, on this line. The image of the real number a is taken to be the directed segment extending from 0 to the point a, *or any other segment having the same length and the same direction.* Conversely, the number a is called the *coordinate* of the vector representing it. Imagine an arrowhead to be marked on the segment at a; and a feather, at 0. Then the vectors representing positive numbers point to the right, and those representing negative numbers point to the left. To the number 0 corresponds the *null vector,* which has no length and no direction.

Whereas the correspondence between the real numbers and

the points of a straight line affords a particularly vivid graphical illustration of the *order* of the real numbers, the representation of these numbers by vectors is better adapted to illustrate the fundamental operations. To *addition* corresponds a suitable *joining* of vectors (cf. p. 10 and §7). The *difference b − a* is represented by the vector which extends from the point *a* to the point *b*. The *multiplication* of *a* by the positive number *b* signifies the *stretching*, in the ratio 1 : *b*, of the vector corresponding to the point *a*. If *b* is negative, the direction of the vector obtained is reversed besides. *Division* is represented in a corresponding manner.

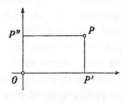

FIGURE 1

No new fundamental considerations are required now to proceed to the foundations of analytic geometry of the plane: We lay down *two* straight lines or *axes* in the plane, of which the second results from the first by rotating the latter about the origin in the *mathematically positive*, i.e., counterclockwise, *sense*, through a right angle. A point *P* of the plane is then uniquely determined by its respective (perpendicular) projections *P'* and *P''* on the first and second axes (cf. Fig. 1), these projections, in turn, are uniquely determined by their respective coordinates *x* and *y*. To every point thus corresponds precisely one *ordered* number-pair (*x, y*), i.e., a pair of numbers whose order of succession must be taken into account,—and, conversely, to every such number-pair corresponds precisely one point.

In this sense, then, the totality of all points of our plane is furnished by the totality of all number pairs (*x, y*), and the latter represented by the former. As is well known, the pair of numbers *x, y* are called the (rectangular) *Cartesian coordinates* of the point represented.

For application it is useful in the plane—to a much greater degree than on the line—to consider, in addition to points, *vectors*, i.e., directed segments. Two directed segments (we again imagine the direction to be indicated by means of arrowhead and feather) are said to represent the *same vector*, if they have the same length and the same direction, disregarding position in the plane in all other respects. Such vectors are denoted by small German letters: $\mathfrak{a}$, $\mathfrak{b}$, ... ; they are called *two-dimensional*, as opposed to the one-dimensional vectors previously introduced on the line.

If a vector $\mathfrak{a}$ is projected on the two axes, we obtain on each of them a (one-dimensional) vector; these are termed the *components* of $\mathfrak{a}$. Each represents (on its axis) a real number; together they are the *coordinates* of $\mathfrak{a}$. To every vector thus corresponds an (ordered) pair of numbers, (x, y). Since, conversely, to every such number pair corresponds a (one-dimensional) vector on each of the axes, and, working backwards, these two can be regarded as the projections of precisely one vector $\mathfrak{a}$ of the plane, we are able to say: The totality of all vectors of the plane is furnished by the totality of all number pairs (x, y), and the latter represented by the former. To the number pair $(0, 0)$ corresponds the *null vector*, which has no length and no direction. If all vectors of the plane are imagined to issue from one and the same point, they are called *coinitial*. If, in particular, they emanate from the origin $(0, 0)$, they are called *radii vectores*. The tip of the radius vector (x, y) then obviously lies just at the point (x, y).

FIGURE 2a   FIGURE 2b

Everyone who is familiar with the parallelogram of forces knows how to join two vectors $\mathfrak{a}$ and $\mathfrak{b}$, i.e., lay the initial point of the second on the terminal point of the first. The vector $\mathfrak{c}$

which extends from the tail of the first to the head of the second (see Fig. 2a) then represents the resultant of the forces signified by the first two vectors.[7] We speak here of *geometric* or *vector addition*, and write, for brevity,

$$\mathfrak{a} + \mathfrak{b} = \mathfrak{c}.$$

If (a, a'), (b, b'), (c, c') are the respective pairs of coordinates of these three vectors, then, as is well known,

$$a + b = c, \qquad a' + b' = c'.$$

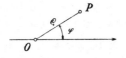

FIGURE 3

In addition to Cartesian coordinates, we use in the plane what are known as *polar coordinates*: A point P of the plane uniquely determines its (non-negative) distance ρ from the origin and, substantially uniquely, the vectorial angle φ of the ray extending from the origin of coordinates, O, toward P; so that P can also be represented by the number pair (ρ, φ) (see Fig. 3). By the vectorial angle of a ray is meant that angle (measured, of course, in radians) through which the direction of the first coordinate axis must be turned in order that it acquire the direction of the ray. Complete rotations in either direction may be neglected here, i.e., arbitrary integral multiples of 2π may be added to, or subtracted from, the angle thus found. The second polar coordinate is thus infinitely multiple-valued, but uniquely determined "mod 2π." That one of the infinitely many values of the same which satisfies the condition $-\pi < \varphi \leqq + \pi$ is called its *principal value*. In general, two angles φ and ψ are already called equal, in symbols: $\varphi = \psi$, if they are congruent mod 2π.

[7]Or, let $\mathfrak{a}$ and $\mathfrak{b}$ issue from the same point; $\mathfrak{c}$, then, is the diagonal of the parallelogram determined by $\mathfrak{a}$ and $\mathfrak{b}$, which emanates from this point (see Fig. 2b).

The polar coordinates ρ, φ and the Cartesian coordinates x, y of one and the same point, distinct from $(0, 0)$, are related by the formulas[8]

(1) $\rho = \sqrt{x^2 + y^2}$, $\cos \varphi = \dfrac{x}{\sqrt{x^2 + y^2}}$, $\sin \varphi = \dfrac{y}{\sqrt{x^2 + y^2}}$,

(2) $\qquad\qquad x = \rho \cos \varphi, \qquad y = \rho \sin \varphi.$

In addition to these few fundamental matters, we shall use only the most familiar facts of analytic geometry of the plane and space, concerning the straight line, the circle, and the sphere, and make use of somewhat more advanced material in applications and examples at the most.

[8]The first of the formulas (1) and the formulas (2) obviously hold also for $(x, y) = (0, 0)$, i.e., for $\rho = 0$ and arbitrary φ.

CHAPTER II

THE SYSTEM OF COMPLEX NUMBERS AND THE GAUSSIAN PLANE OF NUMBERS

4. Historical remarks

The fact that there exists no rational number whose square is equal to 2, that, in other words, the quadratic equation $x^2 - 2 = 0$ has no solution in the system of rational numbers, and many similar facts, have led to the extension of this system to that of the real numbers. For practical application, however, the impossibility mentioned was not of great importance, because there exist rational numbers whose squares are at least nearly (and, indeed, as nearly as one pleases) equal to 2. The situation is entirely different for the equation $x^2 + 2 = 0$ or, say, $x^2 - 10x + 40 = 0$. Here there is no real number x, either rational or irrational, which even "nearly" satisfies the equation. Such deeper impossibilities were noticed early, but *Girolamo Cardano* first made an initial step toward removing them.[9] He is led to consider the last-named equation in connection with the problem of dividing the number 10 into two parts so that their product shall equal 40. He solves it according to the familiar rule which was already generally known at that time, and obtains the (at first quite meaningless) expressions

$$5 + \sqrt{-15} \quad \text{and} \quad 5 - \sqrt{-15}$$

as the two solutions. He notes, however, that *if one operates with these expressions just as with ordinary real numbers*, then, indeed, the sum of the two equals 10 and their product equals 40.

Similar cases were subsequently encountered very frequently; cases, namely, in which one was led, by "formally" correct calculation,[10] to consider expressions containing square roots of negative numbers and yet satisfying, at least "formally," the

[9]G. Cardano, *Artis magnae, sive de Regulis algebraicis liber unus*, . . . , Nuremberg, 1545, ch. 37.

[10]I.e., calculation according to the rules applying to real numbers.

13

conditions of the problem in question. Such expressions were then designated as imaginary, i.e., imagined or unreal, numbers.[11] The most famous example is "Cardan's formulas" for the solution of cubic equations, which, in the case in which the equation possesses three real roots, expresses these roots in the indicated "meaningless" form. Indeed, it turned out, that operation with these "meaningless" expressions could very often produce valuable "real" results: some known ones, by a much shorter route; some new ones; which required considerable time before they could be proved by the customary real method. Often, too, it enabled one to give a more satisfactory form to results already known. One of the most beautiful examples of the latter kind of result is the fundamental theorem of algebra, which asserts that every entire rational function can be expressed as the product of as many factors of the first degree as its degree indicates. Whereas this theorem is not always true (as the above quadratic equation already shows) if one employs only the real numbers, it becomes "formally" correct if one also permits those "meaningless" expressions to appear in factors. An example of the other kind of result is afforded by the expressions for cos nx and sin nx in terms of powers of cos x and sin x (see §11), which are obtained very quickly if one makes use unhesitatingly of roots of negative numbers, but which can be proved by a "purely real" method only in a much more laborious manner. Thus it came about, that roots of negative numbers were not simply rejected, but, on the contrary, were made use of to an ever-increasing extent and with ever greater success; despite the fact that one was unable to assign any direct meaning to them, so that their use remained mysterious and unsatisfying. Most of the things discussed in this little volume were discovered already toward the close of the 17th, and in the course of the 18th, century, especially by *L. Euler* (1707–1783). But not until the turn of the 18th century did one begin to see clearly here. A memoir of the surveyor

[11]This designation has been employed since the middle of the 17th century. As opposed to these numbers, all ordinary numbers were called *real numbers*. Such an opposition of *real* and *imaginary* is found probably for the first time in the famous *Géométrie* of Descartes (Leyden, 1637).

Caspar Wessel, dating from the year 1797, and likewise one of *J. R. Argand,* dating from 1806, which gave a solution of the mystery, at first received no notice. Similar essays of several other mathematicians fared no differently. Only after *C. F. Gauss* developed,[12] in 1831, the same interpretations, independently of his predecessors, had the time become ripe for a full understanding of these things. Within a short time, owing especially to the purely arithmetically treated presentation of *W. R. Hamilton* in the year 1837,—the works of the previously named mathematicians presented matters in geometric garb,— everything mysterious and obscure about these "meaningless expressions" had vanished. Today, due to a clarified attitude toward the foundations of our science, they offer no ideal or actual difficulties whatsoever.

5. Introduction of complex numbers. Notation

The system of real numbers proved to be in many respects, much more efficient than the system of rational numbers, especially in application to geometric questions (see §3): The system of real numbers can be mapped in a one-to-one manner on the points or vectors of a straight line, and the operations on the real numbers can be interpreted as operations on the points or vectors of the straight line. This interpretation urges one, as it were, to attempt, with the new impossibilities discussed in §4 in mind, to define a set of operations for the points and vectors of the plane (see §3), and in this way create a system of elements to which the deficiencies of the system of real numbers no longer adhere. According to §3, such an attempt is equivalent to trying to define a set of operations for number pairs. The first takes place in the language and representations of geometry; the second, in those of arithmetic. In what follows, we shall *always employ both, side by side,* putting the arithmetic interpretation foremost for all fundamental concepts and definitions because of its logical purity, while using the geometric form to facilitate comprehension and the acquirement of a general view through its intuitive power.

[12]C. F. Gauss, Göttingische gelehrte Anzeigen, April 23, 1831; *Werke,* vol. 2, pp. 167–178.

We consider, then, the totality of all ordered pairs made up of two real numbers: (α, α'), (β, β'), ...[13] Visually interpreted, we consider the totality of all points, or that of all vectors, of a plane provided with a pair of coordinate axes in accordance with §3. It will turn out that we shall be able, under suitable agreements as to the meaning of equality and inequality, addition and multiplication, to *operate* with these entities; and in fact, in essentially the very same manner as with real numbers. We shall see, in other words, that these entities can be regarded as *numbers* (see §2, p. 5). Reserving the proof of this for the immediately succeeding paragraphs, we shall now already call them *numbers*, more particularly: *complex numbers*. We denote them by small Roman letters, setting, say,

$$(\alpha, \alpha') = a, (\beta, \beta') = b, \ldots,$$

and shall at the same time employ $a, b, \ldots$ as symbols for the points or vectors of the plane which represent the respective number pairs (α, α'), (β, β'), ... Thus, complex numbers are nothing but ordered pairs of real numbers, or points or vectors of the plane, for which an equality, an addition, and a multiplication have been defined in a definite manner (amplified in §§6–8). The plane in which we imagine these points and vectors to be drawn is called the *plane of complex numbers*, also the *Gaussian number-plane*, or, briefly, the *complex plane*.

For historical reasons, and because of connections which the following paragraphs will reveal more exactly, we designate the first of the two (Cartesian) coordinates of the point a as the *real part*, the second, as the *imaginary part*, of the complex number a, and, accordingly, write

$$(1) \qquad \Re(a) = \alpha, \qquad \Im(a) = \alpha'.$$

We correspondingly designate the first of the two coordinate axes as the *axis of reals*, the second, as the *axis of imaginaries*; and also distinguish their halves into *positive* and *negative half-axes*. Each of the axes divides the plane into two half-planes,

[13]Since we wish to reserve small Roman letters $a, b, \ldots$ for the complex numbers now to be created, real numbers will be denoted by small Greek letters in this and the following paragraphs to §15.

which are called, in virtue of their position, *upper* and *lower*, *left* and *right, half-planes*, respectively. The origin of coordinates, i.e., the point or the number pair (0, 0), or the null vector, is called, briefly, the *point* 0 of the plane, or simply the *origin*. Those complex numbers whose representative points lie on the axis of reals, and whose vectors, consequently, are parallel to it, are called, for brevity, *real*; all the remaining ones, *not real* or *non-real*; those for which the representative points lie on the axis of imaginaries (or whose vectors are parallel to it) are called *pure imaginary.*[14]

The first of the two *polar coordinates* (introduced in accordance with §3) of the point *a*, or, in other words, the length of the vector *a*, which we shall denote by ρ, is called the *modulus* or the *absolute value* of the complex number *a*; the second, φ, which gives the direction of the vector *a*, is called its *amplitude;*[15] in symbols:

(2) $\qquad | a | = \rho, \qquad \text{am } a = \varphi.$

The amplitude of a complex number is thus infinitely multiple-valued, just as the second polar coordinate. All its values, however, differ only by integral multiples of 2π: they are "congruent to one another mod 2π." That value of the amplitude which satisfies the condition $-\pi < \varphi \leqq + \pi$ is called the *principal value* of the amplitude of *a*. Two complex numbers are said to have *the same* amplitude if the two amplitudes are congruent mod 2π, or, in other words, if their principal values coincide. The absolute value of a complex number is a real, non-negative number; it is equal to zero only if the complex number in question is (0, 0), i.e., zero. For this number, the amplitude is regarded as undefined or indeterminate.

The connection between the Cartesian coordinates (α, α') and the polar coordinates ρ, φ of a complex number *a* different from (0, 0) is given, according to §3, by the formulas[16]

[14]As we have already remarked, these last designations will become more understandable through the considerations of §10.

[15]The term *argument* is also in use; in symbols: arg $a = \varphi$.

[16]Cf. p. 12, footnote 8.

$$\rho = \sqrt{\alpha^2 + \alpha'^2}, \qquad \cos \varphi = \frac{\alpha}{\sqrt{\alpha^2 + \alpha'^2}},$$

(3)

$$\sin \varphi = \frac{\alpha'}{\sqrt{\alpha^2 + \alpha'^2}},$$

(4) $$\alpha = \rho \cos \varphi, \qquad \alpha' = \rho \sin \varphi$$

(see Fig. 4).

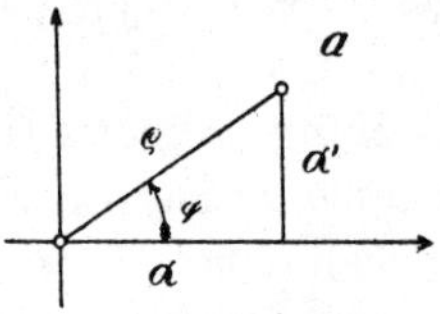

FIGURE 4

If two complex numbers differ only in the sign of the second (Cartesian or polar) coordinate, they are called *complex conjugates*, or simply *conjugates*, of one another. The corresponding points lie symmetric with respect to the axis of reals (see Fig. 5). If one of them is called a, the other is commonly denoted by $\bar{a}$. If they differ in the sign of both Cartesian coordinates, they are called *negatives* of one another. If one of them is called a, the other is denoted by $-a$. The corresponding points lie symmetric with respect to the origin (see Fig. 6); the vectors are parallel and equal in length, but have opposite directions.

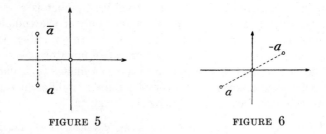

FIGURE 5   FIGURE 6

The task of the next few paragraphs will be, to demonstrate that the entities which we have spoken of here are *numbers* in the sense of §2. The last considerations of this same §2, however, would seem, in principle, to make such a proof impossible. For there we asserted the system of real numbers to be (essentially) the *only* system of entities with which one can operate so that thereby *all* the fundamental laws of arithmetic listed in §2 are valid. This is, of course, true. Yet we shall see that, after making a single minor modification of the fundamental laws, that proof can be furnished. This modification will consist in not demanding any longer, in the fundamental laws of order, that between every pair of our complex numbers always one of the *three* relations $<$, $=$, $>$ hold, but requiring only that there subsist between them one of the *two* relations $=$, $\neq$. The *order* of the complex numbers is thus different in principle from that of the real numbers. We shall then show that under suitable agreements regarding equality, addition, and multiplication, our number pairs obey all the fundamental laws of arithmetic, provided that those laws in which one of the symbols $<$, $>$ appears is modified in the manner now prescribed, or suppressed. It is therefore justified, and also customary, to designate these number pairs, likewise, as *numbers*. To distinguish them from the hitherto existing *real numbers*, however, we call them *complex numbers*.

6. Equality and inequality

The equality of two complex numbers is, naturally, defined by means of the coincidence of the representative points or vectors:

DEFINITION. *The complex numbers $a = (\alpha,\alpha')$ and $b = (\beta,\beta')$ are called equal, in symbols: $a = b$, if simultaneously*

$$\alpha = \beta \qquad and \qquad \alpha' = \beta'.$$

They are, accordingly, called unequal, in symbols: $a \neq b$, if

$$either \qquad \alpha \neq \beta \qquad or \qquad \alpha' \neq \beta'$$

(or both).

The fundamental laws I, 2 to I, 4 obviously are valid in virtue of this definition.[17] I, 1 now reads more simply:

[17]When we use the word "obviously" here and in the following, we mean, of course, that the proof of the assertion is so simple that we may leave it

If a and b are any two numbers, they satisfy precisely one of the relations

(I, 1) $a = b$, $a \neq b$.

Relation I, 5 (the transitivity of the relation $<$) is lost, however; for from $a \neq b$ and $b \neq c$ need not follow $a \neq c$.[18]

7. Addition and subtraction

The way to add two number pairs is suggested by the diagram of the parallelogram of forces:

DEFINITION. *By the sum of two complex numbers $a = (\alpha, \alpha')$ and $b = (\beta, \beta')$ we shall mean the complex number*

$$c = (\alpha + \beta, \alpha' + \beta');$$

in symbols:

$$a + b = c.$$

The formation of the sum of a and b is illustrated graphically (cf. vector addition in §3) *either* by joining the vectors a and b, i.e., placing the initial point of b on the terminal point of a,— the sum c is then the vector which extends from the initial point of a to the terminal point of b (Fig. 7a),—*or* by letting a and b issue from a common point, in which case the sum c is that diagonal of the parallelogram determined by a and b, which emanates from this point (Fig. 7b).

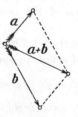

FIGURE 7a   FIGURE 7b

to the reader. It is strongly recommended, however, that he actually carry out carefully each such demonstration.

[18] I, 5 could, at the most, be retained in the form: "*If $a = b$ and $b \neq c$, then $a \neq c$*," but the truth of this assertion is already a consequence (indirectly) of I, 3 and 4.

The fundamental laws II, 1 and 2 are certainly satisfied in view of this definition. That II, 3 and 4 are also satisfied[19] follows immediately from the definition, because these laws hold for each of the two constituents of a number pair taken separately. It follows also just as transparently from the geometric addition of the representative vectors. The validity of law II, 5, which now takes the form:

(II, 5) If $a \neq b$, then invariably $a + c \neq b + c$,—[20]results in an equally simple manner.

Fundamental law III of subtraction is also satisfied. For, given a and b, the complex number $(\beta - \alpha, \beta' - \alpha')$ obviously accomplishes what is required of x in III. Hence, the *difference* $b - a$ is understood to be the number

$$b - a = (\beta - \alpha, \beta' - \alpha').$$

If we allow the vectors a and b to issue from a common point, the difference $b - a$ is represented by the vector extending from the tip of a to the tip of b (Fig. 8a). If we use the points

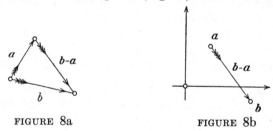

FIGURE 8a   FIGURE 8b

a and b, then $b - a$ is represented by the vector extending from a to b, i.e., from the subtrahend to the minuend (Fig. 8b). In particular,

$$| b - a | \qquad \text{or} \qquad | a - b |,$$

accordingly, means simply the *distance* between the two points a and b. That (now necessarily existent) complex number which,

[19]In II, 4 and 5, c now, of course, denotes an arbitrary, third complex number.

[20]This follows also from II, 2 and III.

when employed as addend in the operation of addition, produces no change, is obviously the number (0, 0). We therefore call it *zero* (cf. what was said in §5), and denote it briefly by 0. The point which represents it is the origin of coordinates, the representative vector is the null vector. Now, further, $0 - a$, or $-a$, signifies the vector which extends from the point a to 0; it therefore has the same length as a, but the opposite direction. Consequently, the complex number $-a$ is the negative of a, which is in agreement with what has already been said in §5.

8. Multiplication and division

It would seem natural to consider, in analogy with the definition of addition, the product of the number pairs (α, α') and (β, β') to be the number pair $(\alpha\beta, \alpha'\beta')$. This multiplication would, in fact, satisfy the fundamental laws IV, 1 to 5, but *not* fundamental law 6, which here must read:

(IV, 6) If $a \neq b$ and $c \neq 0$, then $ac \neq bc$.[21]

Thus, unfortunately, multiplication of number pairs cannot be defined in so simple a manner. The historical development (see §4), however, led early to a definition of multiplication which satisfies all the fundamental laws IV and V, and which can be shown to be essentially the only one possessing this property.

DEFINITION. *The product of two complex numbers* $a = (\alpha, \alpha')$ *and* $b = (\beta, \beta')$ *shall be understood to be the complex number*

$$c = (\alpha\beta - \alpha'\beta', \ \alpha\beta' + \alpha'\beta);$$

in symbols:

$$a \cdot b = c \quad or \quad ab = c.$$

That the multiplication so defined satisfies the fundamental laws IV, 1 to 3, is immediately evident. The laws IV, 4 and 5 are also verified, but the proof of this requires a little calcula-

---

[21]For, take $a = (\alpha, 0)$, $b = (\beta, 0)$, with $\alpha \neq \beta$, and $c = (0, \gamma)$, with $\gamma \neq 0$. Then the products ac and bc formed according to the above rule would both be equal to (0, 0), and hence, to each other. This example shows, at the same time, that, with this definition, the "theorem" to be proved next would be false.

tion. We demonstrate it for IV, 4: Let $c = (\gamma, \gamma')$ be an arbitrary, third complex number. Then

$$(ab)c = (\alpha\beta - \alpha'\beta', \alpha\beta' + \alpha'\beta)(\gamma, \gamma')$$

$$= (\alpha[\beta\gamma - \beta'\gamma'] - \alpha'[\beta\gamma' + \beta'\gamma],$$

$$\alpha[\beta\gamma' + \beta'\gamma] + \alpha'[\beta\gamma - \beta'\gamma']).$$

For $a(bc)$, on the other hand, we find the number pair

$$(\alpha[\beta\gamma - \beta'\gamma'] - \alpha'[\beta\gamma' + \beta'\gamma], \alpha[\beta\gamma' + \beta'\gamma] + \alpha'[\beta\gamma - \beta'\gamma']).$$

Since laws IV hold for the real numbers α, β, . . . , the two number pairs obtained are indeed the same. The proof of IV, 5 is entirely similar, and we leave it to the reader.

It is now easy to see that our multiplication also satisfies law IV, 6 in the modified form[22] (see above): Since it can be stated in the form "If $b - a \neq 0$ and $c \neq 0$, then $(b - a) \cdot c \neq 0$", it is obviously contained in the following

THEOREM. *A product of two complex numbers is equal to zero if, and only if, at least one of the two factors is equal to zero.*

In fact, if $a \neq 0$, but $ab = 0$, then, of necessity, $b = 0$. For, $ab = 0$ means, if $a = (\alpha, \alpha')$ and $b = (\beta, \beta')$, that the two equations

$$\alpha\beta - \alpha'\beta' = 0 \qquad \text{and} \qquad \alpha\beta' + \alpha'\beta = 0$$

hold. If we multiply the first of these by α, the second, by α', and add them, we obtain

$$(\alpha^2 + \alpha'^2)\beta = 0.$$

But since the very meaning of $a \neq 0$ is, that the real number $\alpha^2 + \alpha'^2 \neq 0$, and since our theorem is true for real numbers (see §2, IV, Theorem), it follows that we must have $\beta = 0$. By eliminating β, it is found, in an entirely analogous manner, that $\beta' = 0$ too; and, consequently, $b = 0$.—That, conversely, $ab = 0$ if a or b is equal to 0, follows immediately from the definition of multiplication.

The multiplication specified by the definition thus satisfies all the fundamental laws IV. A calculation similar to the one just

[22]This follows also from IV, 2 and V.

carried out shows, further, that fundamental law V is also fulfilled:

Let $a = (\alpha, \alpha') \neq 0$ and $b = (\beta, \beta')$ be given arbitrarily. Then V asserts the existence of a complex number $x = (\xi, \xi')$ for which $ax = b$, i.e.,

$$(\alpha\xi - \alpha'\xi', \alpha\xi' + \alpha'\xi) = (\beta, \beta'),$$

or

$$\alpha\xi - \alpha'\xi' = \beta \qquad \text{and} \qquad \alpha'\xi + \alpha\xi' = \beta'.$$

Solving these equations simultaneously for ξ and ξ', we obtain, since $\alpha^2 + \alpha'^2 \neq 0$:

$$\xi = \frac{\alpha\beta + \alpha'\beta'}{\alpha^2 + \alpha'^2}, \qquad \xi' = \frac{\alpha\beta' - \alpha'\beta}{\alpha^2 + \alpha'^2}.$$

The number pair x composed of these numbers ξ, ξ' then satisfies law V, and, consequently, furnishes the *quotient*, b/a, of the two complex numbers b and a, provided that $a \neq 0$.

We shall not discuss the geometric representation of multiplication and division until we have become acquainted, in §11, with the trigonometric representation of complex numbers.

9. Derived rules. Powers

Through the considerations of §§6–8, we have now supplied the demonstration, demanded in §5, of the following fact: If the number pairs, or, what amounts to the same, the points or the vectors of the plane, are denoted briefly by small Roman letters, and are called simply numbers; and if, further, equality, addition, and multiplication are defined as we have defined them; then all fundamental laws of arithmetic in which only the equality sign appears remain valid without exception. Of the remaining laws, some become meaningless, some must be slightly altered in form.

From this it follows (cf., in this connection, the details given in §2), quite mechanically, that *all further rules* of the ordinary literal calculus, in which only the equality sign appears, remain valid without further ado, if the letters now denote complex numbers. We say briefly: *We may operate formally with complex numbers the same as with real numbers.* A correct literal calculus "in the real domain," in which only equalities occur, remains valid also "in the complex domain." In operating with inequalities, however, it is to be noted that law I, 5 has dropped out, and that I, 1 as well as II, 5 and IV, 6 must be modified in the manner agreed upon.

As the simplest examples, we stress, above all, the *operation with powers* and the validity of the *binomial theorem*: If a is an arbitrary complex number, and n is a natural number, then the product of n factors, all equal to a, is denoted, as in the real domain, by a^n. And the same considerations as there lead one, in case the *base* $a \neq 0$, to understand by a^0 the number 1, and by a^{-n} the value $1/a^n$. According to these stipulations, the *power* a^k is defined for every integral *exponent* k, and for operating with these powers we have the three rules

$$a^k \cdot a^l = a^{k+l}, \qquad (a^k)^l = a^{kl}, \qquad a^k \cdot b^k = (ab)^k,$$

where the base 0 may appear only if the exponent is positive.

If a and b are two arbitrary complex numbers, and if n is a natural number, then

$$(a + b)^n = a^n + \binom{n}{1}a^{n-1}b + \cdots + \binom{n}{\nu}a^{n-\nu}b^\nu + \cdots + b^n$$

$$= \sum_{\nu=0}^{n} \binom{n}{\nu}a^{n-\nu}b^\nu.$$

The fundamental operations addition, subtraction, multiplication, and division are named the *four rational operations*. If division is excluded, we speak of the *integral rational* operations. An expression which (as, e.g., the preceding one) is formed from any literal quantities and numbers by applying the rational operations (a finite number of times) is therefore called a *rational expression*, and, in particular, an *integral* rational expression, if only the integral rational operations are employed.

10. The system of complex numbers as an extension of the system of real numbers

On the basis of the historical development, which is reflected in the terminology introduced in §5, it has been regarded as a foregone conclusion that the system of complex numbers is an *extension* of the system of real numbers. We have as yet to clarify the extent to which this is the case. For, *pairs* of real numbers are not themselves real numbers. Every number pair $(\alpha, 0)$, however, whose second coordinate is equal to 0, whose representative point consequently lies on the "axis of reals," appears, nevertheless, to be actually a real number in a certain sense. What does this mean? Just this: If every number pair $(\alpha, 0)$ is written in abbreviated form simply as α, then one verifies immediately, that every calculation which proceeds according to the rules laid down for operating with number pairs, and which employs number pairs exclusively of the form $(\alpha, 0)$, goes over into a correct calculation with real numbers. In fact, equality, sum, and product of two such pairs $(\alpha, 0)$ and $(\beta, 0)$ thereby go over, respectively, into equality, sum, and product of α and β; and that suffices. This is expressed by saying: The subsystem of all number pairs $(\alpha, 0)$ is *isomorphic*, with respect to the operations of addition and multiplication, to the system of all real numbers. For this reason we may actually set

$$(\alpha, 0) = \alpha,$$

and, without hesitation, regard the pair $(\alpha, 0)$ as identical with the real number α.

But then we may set $(0, \alpha') = \alpha'(0, 1)$, because, according to §8,

$$\alpha'(0, 1) = (\alpha', 0) \cdot (0, 1) = (0, \alpha').$$

And now an arbitrary number pair can be represented in the form

$$(\alpha, \alpha') = (\alpha, 0) + (0, \alpha') = \alpha + \alpha'(0, 1):$$

All number pairs can thus be represented in this form with the exclusive use of the *single* number pair $(0, 1)$. If, for abbreviation, we replace this number pair by the letter i, as *Euler* first did:[23]

$$(0, 1) = i,$$

[23]In the memoir *De formulis differentialibus* . . . , which was presented to the St. Petersburg Academy in 1777, but which was not published until after Euler's death. A systematic use of the letter i for the imaginary unit was first made by *Gauss*, who availed himself of the same since 1801.

then we can write

$$(\alpha, \alpha') = \alpha + \alpha'i,$$

and this representation is obviously fully unique. Finally, according to §8,

$$(0, 1) \cdot (0, 1) = (-1, 0) = -1,$$

and so

$$i^2 = -1.$$

Thus, our new number system contains numbers whose squares are real and negative;—and likewise, as it will turn out, all the remaining "impossibilities" mentioned in §4 have now turned into realities. In this sense, then, it constitutes a consistent extension of the system of real numbers, and, indeed, one which no longer possesses the deficiencies of the latter.

Since every complex number $a = (\alpha, \alpha')$ can be represented in the form

$$a = \alpha + \alpha'i,$$

the operations with complex numbers can also be regarded as operations with sums of this form, in which α and α' are real numbers, and i is a number symbol for which i^2, i.e., $(0, 1) \cdot (0, 1)$, is equal to -1.

The same goal—the removal of the deficiencies of the system of real numbers by means of suitable extensions of the same, consistent with the fundamental laws—cannot be reached in any (essentially) different manner; but we shall not go into this.

11. Trigonometric representation of complex numbers

In what precedes, we have used Cartesian coordinates to represent points and vectors. If we take polar coordinates, some things become simpler, others, less simple.

If ρ and φ are the polar coordinates of the point $a = (\alpha, \alpha')$, then, according to §5,

$$\alpha = \rho \cos \varphi, \qquad \alpha' = \rho \sin \varphi.$$

The *number* $a = \alpha + \alpha'i$ can therefore be represented in the form

(1) $$a = \rho \cos \varphi + i\rho \sin \varphi = \rho \, (\cos \varphi + i \sin \varphi).$$

This is called the *trigonometric representation of a complex number*. In antithesis to it, the representation $a = \alpha + \alpha'i$ may be designated as the *Cartesian*. In the latter, the real and imaginary parts are displayed; in the former, absolute value and amplitude. The last quantity appears only in the combination (cos $\varphi + i$ sin φ); this factor is called the *direction factor* of the complex number a.

If we have two complex numbers

$$a = \alpha + \alpha'i = \rho(\cos \varphi + i \sin \varphi),$$
$$b = \beta + \beta'i = \sigma(\cos \psi + i \sin \psi),$$

then $a = b$ if, and only if, $\rho = \sigma$ and, at the same time, $\varphi = \psi$, i.e., $\varphi \equiv \psi \pmod{2\pi}$.

The sum and difference of a and b cannot be expressed so simply with the use of the trigonometric representation. The derivation of these expressions is recommended to the reader as an exercise. Since, however, the vectors a, b, and $a + b$ (see Fig. 7a) form a triangle, the well-known theorem that the sum of two sides of a triangle is at least equal to the third, yields the important *inequality*

$$(2) \qquad | a + b | \leqq | a | + | b |,$$

which is called, for brevity, the *triangle inequality*. By the corresponding theorem for the difference of two sides of a triangle, the representation of the difference $b - a$ in Fig. 8a yields the further inequality

$$(3) \qquad | b - a | \geqq | b | - | a |.$$

Multiplication and division, on the other hand, become simpler. First, according to §8,

$$ab = (\alpha\beta - \alpha'\beta') + (\alpha\beta' + \alpha'\beta)i.$$

With the use of polar coordinates we get

$$\alpha\beta - \alpha'\beta' = \rho\sigma(\cos \varphi \cos \psi - \sin \varphi \sin \psi) = \rho\sigma \cos (\varphi + \psi),$$
$$\alpha\beta' + \alpha'\beta = \rho\sigma(\cos \varphi \sin \psi + \sin \varphi \cos \psi) = \rho\sigma \sin (\varphi + \psi),$$

so that

(4) $$ab = \rho\sigma \left[\cos (\varphi + \psi) + i \sin (\varphi + \psi)\right].$$

Thus, the absolute value of this product is equal to the product of the absolute values of the factors, the amplitude of the product is equal to the sum of the amplitudes of the factors; in symbols:

(5) $$|\, ab \,| = |\, a \,| \cdot |\, b \,|, \qquad \text{am } ab = \text{am } a + \text{am } b.$$

It is shown in an entirely similar manner (or by starting from $(b/a)\cdot a = b$, $a \neq 0$), that, for $a \neq 0$,

(6) $$\frac{b}{a} = \frac{\sigma}{\rho} \left[\cos (\psi - \varphi) + i \sin (\psi - \varphi)\right],$$

(7) $$\left|\, \frac{b}{a} \,\right| = \frac{|\, b \,|}{|\, a \,|}, \qquad \text{am } \frac{b}{a} = \text{am } b - \text{am } a.$$

Since $|\, 1 \,| = 1$, and we can set $\text{am } 1 = 0$, we have, in particular,

(8) $$\left|\, \frac{1}{a} \,\right| = \frac{1}{|\, a \,|}, \qquad \text{am } \frac{1}{a} = -\text{am } a,$$

(9) $$\frac{1}{a} = \frac{1}{|\, a \,|} (\cos \varphi - i \sin \varphi).$$

Repeated application of multiplication and division leads, finally, to what is known as *de Moivre's formula*:

(10) $$a^n = [\rho(\cos \varphi + i \sin \varphi)]^n = \rho^n(\cos n\varphi + i \sin n\varphi),$$

where n may be an arbitrary integer. From it we get, in particular,

(11) $$|\, a^n \,| = |\, a \,|^n, \qquad \text{am } (a^n) = n(\text{am } a).[24]$$

If we make use of the binomial theorem, there follows, for positive integral n:

$$\cos n\varphi + i \sin n\varphi = \cos^n\varphi + \binom{n}{1}i \cos^{n-1}\varphi \sin \varphi$$

$$- \binom{n}{2} \cos^{n-2}\varphi \sin^2\varphi - \binom{n}{3}i \cos^{n-3}\varphi \sin^3\varphi + + - - \cdots .$$

[24] The last is true in the sense that every value on the right is contained among the values on the left; but in general there are more values on the left than on the right.

If we now separate real and imaginary parts, we obtain the representation, mentioned in §4, of cos $n\varphi$ and sin $n\varphi$ in terms of powers of cos φ and sin φ.

12. *Geometric representation of multiplication and division*

From the considerations of the preceding paragraph, there follows, now, the very simple representation of multiplication and division. If

$$a = \rho(\cos \varphi + i \sin \varphi) \quad \text{and} \quad b = \sigma(\cos \psi + i \sin \psi)$$

issue from a common point, say 0, then from formulas (4) and (6) there we read off the following constructions of the vectors ab and b/a:

Rotate the vector b through the angle φ = am a in the positive sense,[25] *and stretch it in the ratio* $1 : \rho = 1 : |a|$.[26] *The new vector represents the product ab.*

If the vector b is rotated through the angle φ in the negative sense (i.e., through the angle $-\varphi$ in the positive sense), and stretched in the ratio $\rho : 1$, we obtain the vector b/a.

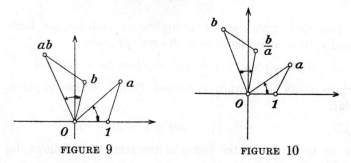

FIGURE 9   FIGURE 10

[25]When we speak of a *rotation through the angle φ in the positive sense*, here and in what follows, this will invariably mean that the rotation is to be made through the angle $|\varphi|$ in the *positive or negative* sense, according as the number φ is positive or negative.

[26]When we speak of a *stretching* in the ratio $\gamma : \delta$ (γ, δ real and positive), this will always mean that the length of the old segment is to that of the new, as $\gamma : \delta$,—irrespective of the relation $>$, $=$, $<$ in which γ stands to δ. Thus, the word *stretching* is also employed when an actual shrinking is intended, or even no change at all.

The following is merely a somewhat different form of this construction:

Plot (see Fig. 9) the *points* 0, 1, *a*, and *b*. On the segment extending from 0 to *b*, place a triangle which is directly similar to the triangle 01*a*, in such a manner, that the segments 0 . . . 1 and 0 . . . *b* become corresponding sides. Then the *third* vertex of this triangle is the point *ab*.

If, however, we attach the triangle to 0 . . . *b* so that the sides 0 . . . *a* and 0 . . .*b* correspond to one another, then the *third* vertex represents the quotient *b/a*. In particular, we find the point 1/*a*, the *reciprocal* of *a*, by placing on the segment 0 . . . 1 a triangle directly similar to 01*a*, in such a manner, that the segments 0 . . . *a* and 0 . . . 1 become corresponding sides. The third vertex of this triangle furnishes the point 1/*a* (see Fig. 11).

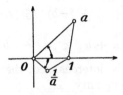

FIGURE 11

13. Inequalities and absolute values. Examples

Of the fundamental laws pertaining to inequalities, I, 5 has dropped out, and the remaining three have acquired the simplified forms stated in §§6, 7, and 8.

These rules are so simple, that their application requires no further explanation. On the other hand, we shall go somewhat more closely into the manipulation of *absolute values*, since it is made use of especially often in what follows. It rests essentially on the three facts ascertained in the preceding paragraphs:

I. *The absolute value* | *a* | *of a complex number a is a real, non-negative number, which is equal to* 0 *if, and only if, a* = 0.

II. $\qquad\qquad | ab | = | a | \cdot | b |.$

III. $\qquad\qquad | a + b | \leq | a | + | b |.$

The proof of the last fact, the so-called *triangle inequality*, was carried out geometrically. It can be given analytically, as follows: The assertion is that for four arbitrary real numbers α, α', β, β' we have invariably

$$\sqrt{(\alpha + \beta)^2 + (\alpha' + \beta')^2} \leqq \sqrt{\alpha^2 + \alpha'^2} + \sqrt{\beta^2 + \beta'^2}.$$

If we square twice, we find that this inequality is certainly correct if the inequality

(1) $$(\alpha\beta' - \alpha'\beta)^2 \geqq 0$$

holds,—and this last is surely the case.

From II, or from the definition itself, it follows that

$$|-a| = |a|;$$

and then from III we obtain, further,

$$|a| = |(a + b) + (-b)| \leqq |a + b| + |b|$$

or

$$|a + b| \geqq |a| - |b|.$$

Since a and b may be interchanged, there results, finally, the somewhat sharper inequality

III'. $$|a + b| \geqq ||a| - |b||,$$

to which the name *triangle inequality* is likewise applied.

We close this chapter with the presentation of a series of simple applications which result from the agreements and theorems of the preceding paragraphs and which are often used in what follows.

1. From the triangle inequality, by repeated application, follows: If a_1, a_2, ..., a_p are any p complex numbers, then

$$|a_1 + a_2 + \cdots + a_p| \leqq |a_1| + |a_2| + \cdots + |a_p|.$$

The vector $(a_1 + a_2 + \cdots + a_p)$ is found by joining the vectors a_1, a_2, ..., a_p in succession and then connecting the initial point of the first with the terminal point of the last. For the product, however, we have

$$|a_1 a_2 \ldots a_p| = |a_1| \cdot |a_2| \ldots |a_p|.$$

Since the absolute value $|z|$ of a complex number[27] z represents its distance from the origin, and the absolute value $|z_2 - z_1|$ of the difference of two numbers z_1 and z_2 represents the distance between the corresponding points, the following simple facts ensue:

2. If, for a complex number z, the absolute value $|z| = 1$, then its image point lies on the so-called *unit circle*, i.e., the circle with radius 1 about the origin as center. Conversely, $|z| = 1$ for every point z on this circle. In this sense the equation $|z| = 1$ can be regarded as the *equation of the unit circle*: It is satisfied if, and only if, z lies on the unit circle. For these, and only these, z, the trigonometric representation has the form

$$z = \cos \varphi + i \sin \varphi.$$

3. Invariably (i.e., for every complex number z),

$$|\Re(z)| \leq |z| \qquad \text{and} \qquad |\Im(z)| \leq |z|.$$

For, the legs of a right triangle are not greater than its hypotenuse.

4. As under 2., if a is a given complex number, and ρ denotes a real and positive number, then

$$|z - a| = \rho$$

is the equation of the circle with radius ρ about the point a as center: A number z satisfies this equation if, and only if, its image point lies on the circle in question.

5. Similarly, $|z - a| \leq \rho$, or $|z - a| < \rho$, is the equation of the *surface* of this circle. In the first case, the circumference, the *boundary* of the circular surface, is counted with the latter, in the second, not. Inequalities such as

$$|z - a| \geq \rho, \qquad |z - a| > \rho, \qquad \rho_2 < |z - a| < \rho_1$$

have an analogous, simple meaning. The last represents the surface of the ring between the circles with radii ρ_1 and ρ_2 about the point a as center, exclusive of the two boundaries.

[27]In the following, the letter z is often used to denote an arbitrary complex number, while the letters $a, b, \ldots$ are reserved for definitely chosen numbers.

6. If ϵ is a given positive number, the circular surface $|z - a| < \epsilon$ is called, briefly, a (circular) ϵ-*neighborhood* of a. If we set $a = \alpha + i\alpha'$, $z = x + iy$, then the square surface determined by

$$|x - \alpha| < \epsilon, \qquad |y - \alpha'| < \epsilon$$

is correspondingly called a *square* ϵ-neighborhood of a.

7. Suppose z is a complex number such that

$$\left| \frac{z - 1}{z + 1} \right| = 1.$$

This means that its distance $|z - 1|$ from the point $+1$ is equal to its distance $|z - (-1)|$ from the point -1. Hence, z lies on the axis of imaginaries. The equation written down can therefore be regarded as the equation of this axis. This is also true, however, of the simpler equation

$$\Re(z) = 0.$$

8. In the same sense, $\Re(z) \geqq 0$ characterizes the right half-plane (including its boundary, the axis of imaginaries); and the interpretation of the inequalities $\Re(z) < 0$, $\Im(z) \leqq 0$, $\Im(z) > 0$ is equally simple.

CHAPTER III

THE RIEMANN SPHERE OF NUMBERS

14. The stereographic projection

Up to now we have used the plane of analytic geometry for graphical illustration of the complex numbers. For many purposes it proves to be more advantageous to employ the sphere to this end. If it is to perform the same function, we must bring about a correspondence between the points of the sphere and those of the plane; must, as we say briefly, map the sphere (in a one-to-one manner) on the plane. The realization of such a mapping is the ancient problem of constructing geographical maps. It can be accomplished in the most varied ways. But it is known that *distortions* are inevitably introduced in the process: It is not possible to carry out the mapping so that the map is geometrically similar to the original. We may, therefore, only inquire, with regard to a given map: Which entities (distances, angles, areas, forms, etc.) bear a fixed ratio to those of the original, which do not, and what is the nature of the changes in the latter case? For our purposes, only the mapping known as the *stereographic projection* comes into question. It is realized as follows:

On the xy-plane of analytic geometry[28] we place a sphere of diameter 1, in such a manner, that it touches the plane at the origin of coordinates, 0. So that we may be able to express ourselves conveniently, we make use of the usual geographical terminology on the sphere, and, accordingly, call the point of contact, O, the south pole, and the diametrically opposite point, the north pole N. We now consider the rays, issuing from this north pole, which intersect the plane, and consequently also intersect the sphere in a second point (distinct from N). *We associate the point (distinct from N) on the sphere with that point*

[28]In this paragraph, all numbers are again supposed to be real; we operate exclusively in the real domain.—We imagine the xy-plane to lie horizontally before us.

of the plane, which lies on the same ray (see Fig. 12). Obviously, to every point P of the plane corresponds, in virtue of this association, precisely one point P' of the sphere, distinct from N; and conversely. In short: The surface of the sphere (from which one must imagine the north pole to be deleted) is mapped in a one-to-one manner on the plane. It is clear that the parts of the sphere lying in the neighborhood of the south pole receive, hereby, only a slight distortion, whereas those situated near the north pole undergo a violent distortion. What does this mapping preserve? It is the purpose of the considerations which follow, to show that the mapping is *circular* and *isogonal*. The first means that every circle on the sphere is mapped into a circle *or a straight line* of the plane (and conversely); the second, that any two circles, and, more generally, any two curves, on the sphere intersect at the same angle as their images in the plane (and conversely).

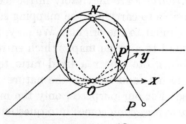

FIGURE 12

We see immediately that the parallels of latitude on the sphere go over into the concentric circles about 0 as center in the plane, while the semimeridians of the sphere correspond to the rays emanating from 0. In particular, since we have given the *diameter* 1 to the sphere, its equator goes over into the circle with radius 1 about 0 as center. For specifying geographical longitude, we shall take that semimeridian to be prime meridian, which corresponds to the positive axis of reals; and, in general, the geographical longitude of a definite semimeridian will be taken to be the vectorial angle of its corresponding ray emanating from 0.

It is immediately evident, furthermore, that a straight line of the plane goes over into a circle through the north pole. For, the rays extending from N toward the points of the straight line form a half-plane which cuts the sphere in the image circle of the straight line. We now show: *Two straight lines in the plane intersect at the same angle as their image circles on the sphere.* This is almost obvious. For if we have two straight lines in the plane, which intersect at the point P, then their image circles intersect at the point P' which corresponds to P, *and at the north pole, N,* and certainly at the same angle at both points. If we draw the tangents to the two image circles at the north pole, then these are parallel (in space) to the given lines, because the sphere's tangent plane at the north pole is parallel to our xy-plane. The angle between the straight lines at P in the plane is consequently the same as the angle between the two image circles at the north pole, and hence, is also the same as that between the image circles at P'.

Suppose we have a curve $\Re$ in the plane, with a tangent to it at P. Then, under the mapping on the sphere, this figure goes over into a certain image curve $\Re'$ and the image circle of the tangent, which circle is tangent to $\Re'$ at the image point, P', of P. And from this it then follows immediately that any two curves in the plane intersect at the same angle as their image curves on the sphere: *The stereographic mapping is isogonal.*

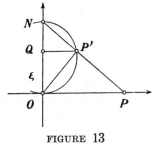

FIGURE 13

In order to establish the *circularity* of the mapping, we shall represent it analytically. To this end, we introduce a rectangular coordinate system $\xi\eta\zeta$ in space, whose ξ- and η-axes coincide, respectively, with the x- and y-axes of our xy-plane, and whose positive ζ-axis has the direction of the diameter ON. If we now pass the plane containing the points O, N, P, P', through the (spatial) Figure 12, we obtain Figure 13, in which we have also drawn the segment OP' and the perpendicular $P'Q$ dropped from P' to ON. In this diagram, $OP = \rho = \sqrt{x^2 + y^2}$ and $OQ = \zeta$. If, for abbreviation, we set $P'Q = \rho'$, the right triangles in the figure yield the two proportions

$$\zeta : \rho' = \rho : 1 \qquad \text{and} \qquad \rho' : (1 - \zeta) = \rho : 1.$$

From these we get

$$\rho' = \frac{\rho}{1 + \rho^2}, \qquad \zeta = \frac{\rho^2}{1 + \rho^2}.$$

If we again denote the vectorial angle of the ray OP in the plane by φ, we have $x = \rho \cos \varphi$, $y = \rho \sin \varphi$ and likewise $\xi = \rho' \cos \varphi$, $\eta = \rho' \sin \varphi$. Putting these results together, we have in

$$(1) \quad \xi = \frac{x}{1 + x^2 + y^2}, \quad \eta = \frac{y}{1 + x^2 + y^2}, \quad \zeta = \frac{x^2 + y^2}{1 + x^2 + y^2}$$

the formulas which lead us from the coordinates x, y of a point P of the plane to the spatial coordinates ξ, η, ζ of the image point P'. From (1) we immediately obtain

$$(2) \qquad x = \frac{\xi}{1 - \zeta}, \qquad y = \frac{\eta}{1 - \zeta}, \qquad \rho^2 = \frac{\zeta}{1 - \zeta},$$

$$(\zeta \neq 1, \text{ i.e., } (\xi, \eta, \zeta) \neq N),$$

for the inverse connection.

If we now consider a circle or a straight line in the plane, this means that we fix our attention on all points (x, y) of the plane, for which an equation of the form

$$(3) \qquad \alpha(x^2 + y^2) + \beta x + \gamma y + \delta = 0$$

is satisfied, where α, β, γ, δ are real numbers. It is a circle or a straight line, according as α is not, or is, equal to 0. For the image points (ξ, η, ζ) on the sphere we have, then, the equation

$$\alpha\zeta + \beta\xi + \gamma\eta + \delta(1 - \zeta) = 0.$$

But this equation is linear. Hence, all our image points lie on the same plane cutting the sphere, and, consequently, form a circle, Q.E.D.

15. The Riemann sphere of numbers. The point ∞. Examples

Let us return to the system of complex numbers. Thus far we have represented the complex number $z = (x, y) = x + iy$ graphically by the point (x, y). We now associate with it also that point on the sphere which corresponds to the point (x, y) by virtue of the stereographic projection described in §14, and call it, too, the *point z*, for brevity. Then the totality of complex numbers corresponds, in a one-to-one manner, to the points, different from the north pole, on the sphere, which is therefore called the *sphere of complex numbers*, the *Riemann sphere*, or, briefly, the *complex sphere*.

It is useful to acquire as vivid a picture as possible of the distribution of the numbers on the complex sphere. We therefore make the following observations, whose justification the reader himself will be able to supply:

1. The unit circle goes over into the equator; the interior of the unit circle corresponds to the southern hemisphere, the exterior, to the northern hemisphere.

2. The rays issuing from O correspond to the semimeridians; the vectorial angle of a ray is the geographical longitude of the corresponding semimeridian. In particular, the positive axis of reals goes over into the prime meridian; the negative axis of reals, into the semimeridian of longitude 180°; the positive and negative halves of the axis of imaginaries, into the semimeridians of longitude ±90°. The circles in the plane about 0 as center correspond to the parallels of latitude on the sphere: if the circle in the plane has the radius r, then the geographical latitude β of the corresponding parallel of latitude[29] is given by the formula

$$(1) \qquad \cot\left(\frac{\pi}{4} - \frac{\beta}{2}\right) = r.$$

---

[29]Northern latitude is reckoned positive; southern, negative. β thus satisfies the condition $-(\pi/2) \leqq \beta < +(\pi/2)$.

This is inferred from Fig. 13, where

$$\measuredangle\ NOP' = \measuredangle\ NPO = \frac{1}{2}\left(\frac{\pi}{2} - \beta\right).$$

The upper half-plane yields the posterior (eastern) hemisphere; the lower half-plane, the anterior (western) hemisphere. The right and left half-planes likewise yield the right and left hemispheres, respectively.

3. According to what precedes, the point $z = (x, y) = x + iy = \rho\ (\cos\varphi + i\sin\varphi)$ yields that point on the sphere with geographical longitude φ and latitude β which is obtained from (1). The points $1, i, -1, -i$ of the sphere lie on the equator and possess the respective longitudes $0°, 90°, 180°, -90°$. Two conjugate numbers z and $\bar{z}$ correspond to two points on the sphere which are symmetric with respect to the plane of the prime meridian. A reflection in the axis of reals in the complex plane corresponds to a reflection in the plane of the prime meridian on the complex sphere.

4. To the pencil of all straight lines through a point P corresponds the pencil of all circles which pass through the image point P' and the north pole; to the pencil of all circles through two points, the pencil of all circles through the two image points on the sphere. To a family of parallel straight lines corresponds a family of circles through the north pole, which have there a common tangent parallel to the straight lines of the plane.

5. A great circle on the sphere is the image of a circle which meets the unit circle of the plane in two diametrically opposite points; and, conversely, every such circle has as its image a great circle on the sphere.

We shall see, in what follows, that some things can be visualized better in the plane, whereas others are easier to see on the sphere. It is better to follow the *four operations on pairs of numbers* (addition, subtraction, multiplication, and division) in the plane. Anything that takes place in the very remote parts of the plane, however, is viewed better on the sphere, because such parts of the plane are mapped on parts of the surface of the sphere which lie in the neighborhood of the north pole.

This situation makes it seem appropriate to regard also the north pole itself as the image of an (improper) point of the plane: the *point* ∞ (*infinity*). We thus add to the system of complex numbers a *single* improper element, the value *"infinity"* (in symbols: ∞); and, accordingly, the Gaussian plane is closed by the *point* ∞. The complex plane which has been closed in this way is then mapped on the full sphere in a one-to-one manner without exception. The hitherto existing points are called *proper points*, to distinguish them from the improper point ∞, and the hitherto existing plane is called the *proper plane*. The turn of expression often used in geometric discourse, that a straight line in the plane may be regarded as "closed at infinity," receives a direct intuitive meaning through the mapping on the sphere, because to the straight line corresponds a circle through the north pole. Many other things, too, will, in like manner, become pictorially clearer on the sphere than in the plane.

CHAPTER IV

MAPPING BY MEANS OF LINEAR FUNCTIONS

16. *Mapping by means of entire linear functions*

The concept of a function will occupy us more intensively in section IV. We speak of a *function,* if to every complex number z there is made to correspond, by means of some rule, a new complex number w. In this section we shall deal with only a very simple correspondence of this kind: If a, b, c, d are definitely given complex numbers, then to every value z there shall correspond the value

$$(1) \qquad\qquad w = \frac{az + b}{cz + d}.$$

In this case we speak of a *linear function.* In the real domain, the behavior of a function $y = f(x)$ is visualized by drawing, in an xy-plane, the corresponding curve whose equation is $y = f(x)$. In the complex domain, *two* planes are required. In one of them, the *z-plane,* we plot the value of the independent variable z; in the other, the *w-plane* or *image plane,* the corresponding value w. If we imagine this to be carried out for all values z, we obtain a mapping of the z-plane on the w-plane. If Riemann spheres are used instead of the planes, we obtain a mapping of the *z-sphere* on the *w-sphere.* Sometimes it is convenient to think of the two planes or the two spheres in question as coincident. We then speak of a mapping of the z-plane, or of the z-sphere, *on itself.* This mode of representation is advantageous in connection with the very first and simplest mappings to be considered.

42

In the next few paragraphs we shall investigate more closely the mappings effected by the linear functions (1). They are called, briefly, *linear mappings* or *linear transformations*. We begin with the *entire linear functions*, i.e., those of the form

$$(2) \qquad\qquad w = az + b.$$

1. Let $a = 1$. We have, then, the function

$$(3) \qquad\qquad w = z + b$$

before us. If $b = 0$, then (3) is the *identity*, in which case image and object coincide. If $b \neq 0$, we obtain the image point of any z by adding the vector b to the radius vector z. The terminal point of b then furnishes the image point w. Thus, from any figure of the plane we obtain the image figure by subjecting the original to the *translation* or *parallel displacement* (b), i.e., the parallel displacement determined in magnitude and direction by the vector b. Image and original are *congruent* to each other.

2. Now let $b = 0$, i.e., let the function be of the form

$$(4) \qquad\qquad w = az,$$

and take $a \neq 0$.[30] Then the image point of any z is obtained by multiplying it by one and the same number a. Thus, according to §12, it is obtained by rotating the radius vector z through the angle am a in the positive sense, and then stretching it in the ratio $1 : |a|$. This mapping is consequently designated, briefly, as the *rotary stretching* (a) with the center 0. If, in particular, $|a| = 1$, so that a is of the form $a = \cos \alpha + i \sin \alpha$, the mapping reduces to a *pure rotation* (α), i.e., a rotation through the angle α, and having 0 as center. This mapping is obviously again a *congruence* mapping. If, on the other hand, $\alpha \equiv 0 \pmod{2\pi}$, i.e., if a is real and positive, say equal to A, then we are dealing with the *pure stretching* $1 : A$. This mapping is a *similarity transformation* with 0 as center; original and image are similar to each other, and their ratio of similitude is $1 : A$.

[30]If $a = 0$, the function is *identically constant*, viz., 0, which means that to every point z, one and the same point is assigned as image. This *degenerate mapping*, which is of no interest, will be disregarded in the future.

For arbitrary $a \neq 0$, the mapping $w = az$ is thus a *similarity mapping* with 0 as center; the ratio of similitude of original to image is $1 : |a|$. Of course, for $a = 1$ it goes over again into the identity.

The rotary stretching (a) can also be carried out by *first* stretching in the ratio $1 : |a|$ and *then* rotating through the angle am a: Rotating and stretching are (if both have the same center) *commutative* operations.

3. If, finally, an arbitrary entire linear function

$$w = az + b$$

is given, where it is assumed merely that $a \neq 0$ (since otherwise the mapping would again degenerate), the image, w, can be obtained from the original, z, by performing first the rotary stretching (a) and then the translation (b). If we write the function in the form

$$w = a\left(z + \frac{b}{a}\right),$$

it is evident that the same end is attained by performing first the translation (b/a) and then the rotary stretching (a),—the latter having that point as center, which is carried into 0 by the translation. Both ways show that the mapping (2) by means of an entire linear function is a *similarity transformation*: Original and image of every figure are similar to each other in the ratio $1 : |a|$.

To visualize this mapping we have used the plane. The sphere is not so well suited for this purpose, because addition (translation) is not represented so vividly on it. Multiplication (rotary stretching) cannot be visualized so well either, though somewhat more clearly: To a rotation of the plane with 0 as center corresponds, naturally, a rotation of the sphere about the north-south axis. In order to picture the stretching $1 : A$ (with 0 as center), one must imagine the surface of the sphere to be drawn (like a rubber membrane) away from the south pole and pushed toward the north pole, or conversely, according as $A \gtrless 1$.

17. *Mapping by means of the function $w = 1/z$*

The mapping by means of the simplest *fractional* linear function

$$(1) \qquad\qquad w = \frac{1}{z}$$

is investigated most easily if we employ polar coordinates in both planes. We set

$$z = \rho(\cos \varphi + i \sin \varphi), \qquad w = \sigma(\cos \psi + i \sin \psi).$$

Then, according to §11, (8),

$$(2) \qquad\qquad \sigma = 1/\rho \quad \text{and} \quad \psi = -\varphi.$$

w and z thus have *reciprocal* absolute values, and amplitudes of *opposite* sign. The transition from z to w is therefore conveniently accomplished in two steps:

1) the passage from a point z to that point z' which has the *same* amplitude but the *reciprocal* absolute value; and

2) the passage from the point z' thus obtained, to that point which has the *same* absolute value but the *negative* amplitude.

The *second* step is particularly easy to survey. It signifies the passage from a number to its conjugate; hence, simply a reflection in the axis of reals in the plane, or, a reflection in the plane of the prime meridian on the sphere.

The *first* step, the passage from the point z with the polar coordinates ρ, φ to that point z' whose polar coordinates ρ', φ' satisfy the equations

$$(3) \qquad\qquad \rho' = 1/\rho \quad \text{and} \quad \varphi' = \varphi,$$

is effected most clearly on the sphere: The points z and z' have the same geographical longitude, and for their respective latitudes β and β' we have, according to §15, 1,

$$\cot \left(\frac{\pi}{4} - \frac{\beta}{2}\right) = \rho, \qquad \cot \left(\frac{\pi}{4} - \frac{\beta'}{2}\right) = \rho'.$$

The two angles appearing in the left-hand members are complementary angles, because $\rho\rho' = 1$; consequently,

(4) $\beta + \beta' = 0, \qquad \beta' = -\beta.$

The geographical latitudes of the points z and z' thus differ only in sign; each of these points can be obtained from the other by means of a *reflection in the equator*.

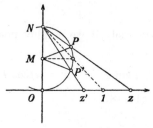

If we regard the plane of the meridian containing both points z and z', as the plane of Fig. 14, we easily infer from the latter, that if the geographical latitudes of P and P' differ only in sign, then we have for their respective images z and z' the relation $\rho\rho' = 1$, *and conversely*.

On the sphere, then, we get from a point z to the point w determined by (1), by reflecting first in the plane of the prime meridian, and then in the plane of the equator. These two reflections obviously can be replaced, however, by a *single* reflection in the line of intersection of these two planes; or, what amounts to the same thing, by a *rotation of the sphere* through 180° about this straight line as axis. This axis connects the points $+1$ and -1 on the sphere.

Thus, the mapping (1), when interpreted on the sphere, is a perfectly clear *congruence* mapping. Since the rotation mentioned carries the south pole into the north pole, and the latter into the former, it is reasonable to regard the points 0 and ∞ as images of each other in virtue of the mapping (1). In this sense (*but also only in this sense*), one sets, in the theory of functions,

(5) $\dfrac{1}{0} = \infty, \qquad \dfrac{1}{\infty} = 0,$

which is to say no more and no less than that, in the mapping (1), 0 goes over into ∞, ∞ into 0. On the basis of this agreement, $w = 1/z$ now effects a one-to-one mapping of the two full spheres on each other. Every point thereby goes over into a definite other point, with the exception of the two points ±1, each of which corresponds to itself. They are called the *fixed points* of the mapping.

In the plane, the interpretation of our mapping is not quite so simple; nevertheless, it is important to be well acquainted with it here, too. The *second* of the above steps was seen to be a reflection in the axis of reals. Every figure is therefore transformed into a congruent one, but "with *reversion of angles*," since, under the reflection, a positive rotation goes over into a negative one, and conversely. The *first* step, which is given analytically by equation (3), requires the passage from a point $z \neq 0$ in the plane, to that point z' which lies on the same ray issuing from 0, but which possesses the reciprocal distance from 0. This mapping, taken by itself, is called the *mapping by*

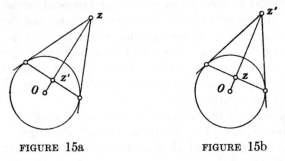

FIGURE 15a   FIGURE 15b

reciprocal radii, reflection in the unit circle, or *inversion with respect to the unit circle.* Its most important properties are the following:

1. $z' = 1/\bar{z}$, because z' is conjugate to $1/z$.[31]

2. Reflection in the unit circle is *involutoric.* That is to say: if z' is the image of z, then, conversely, z is the image of z'.

[31]On the sphere, the point $-z'$ lies diametrically opposite the point z. The points (numbers) z and $-1/\bar{z}$ are therefore said to be *diametrically opposite* each other, or *antipodal.*

3. How to find the reflection, z', in the unit circle, of a point z, by means of elementary geometric construction, can be read off from Figs. 15a and b, in which the circle represents the unit circle in the plane. If z lies in the exterior of this circle, then z' lies in its interior, and conversely. If z lies on the circumference of the unit circle, then z' is identical with z. If z lies very close to the origin, then z' lies far away from the origin.

This, too, makes it understandable why the image of the point 0 is to be regarded significantly as the point ∞, and conversely. *And, above all, it is now perfectly clear why one closes the complex plane with precisely one improper point, just the point* ∞.

The further properties of the mapping by reciprocal radii follow very simply if we pass from the previously used sphere, by stereographic projection, to the plane. Thus, we obtain immediately:

4. *Reflection in the unit circle is circular and isogonal,*—the last, however, *with reversion of angles.* For, the mapping certainly possesses these properties when interpreted on the sphere as an ordinary reflection in the equatorial plane, and then both are preserved by the stereographic projection. Let us distinguish between "true circles" and straight lines in the plane (we regarded them, in §14, as forming a single totality). Then the following special facts concerning *reflection in the unit circle* result without further ado:

a) A straight line which does not pass through the origin becomes a true circle which passes through the origin.[32]

b) A straight line which goes through the origin corresponds to itself as a whole.

c) A true circle which passes through the origin becomes a straight line which does not pass through the origin.

d) A true circle which does not pass through the origin becomes again a true circle which does not pass through the origin.

---

[32] A circle on the sphere, which passes through the north pole but not through the south pole, goes over, under a reflection in the equatorial plane, into a circle on the sphere, which passes through the south pole but not through the north pole.—The proofs of b), c), and d) are of analogous simplicity.

e) Every circle which cuts the unit circle at right angles (such a circle is said to be *orthogonal* to the unit circle) goes over into itself as a whole. (For, corresponding to it on the sphere is a circle which, because of the isogonality and circularity of the stereographic projection, is symmetric with respect to the equator, and, consequently, goes over into itself under a reflection in the equatorial plane.)

f) If two circles which are orthogonal to the unit circle intersect (the straight lines through 0 may also be classed with such circles), then the points of intersection are symmetric with respect to the unit circle. (For, on the sphere, they are symmetric with respect to the equator.) And, conversely, a circle which passes through two points symmetric with respect to the unit circle, is orthogonal to the unit circle.

We have derived the properties of reflection in the unit circle by transferring the corresponding ordinary reflection in the equatorial plane on the sphere, by means of stereographic projection, to the plane. It is not difficult, either, to obtain it without the use of the sphere. Let us denote the Cartesian coordinates of z and z' by (x, y) and (x', y'), respectively. Then $x/\rho = x'/\rho' = \cos \varphi$ and $y/\rho = y'/\rho' = \sin \varphi$. Since $\rho\rho' = 1$, $\rho^2 = x^2 + y^2$, and $\rho'^2 = x'^2 + y'^2$, we have

(6) $$x' = \frac{x}{x^2 + y^2}, \qquad y' = \frac{y}{x^2 + y^2}$$

and

(7) $$x = \frac{x'}{x'^2 + y'^2}, \qquad y = \frac{y'}{x'^2 + y'^2}.$$

Now, every straight line and every circle in the plane can be represented by an equation of the form

$$\alpha(x^2 + y^2) + \beta x + \gamma y + \delta = 0,$$

where α, β, γ, δ denote suitable real numbers. If we replace x, y by their respective values in (7), we find that the image points (x', y') satisfy the equation

$$\alpha + \beta x' + \gamma y' + \delta(x'^2 + y'^2) = 0,$$

and, consequently, again lie on a circle or a straight line: Reflection in the unit circle is a circular transformation. Its isogonality can also be proved without great difficulty by remaining in the plane, but we may leave this to the reader.

It is now clear how to define *reflection in an arbitrary circle*: Let $\mathfrak{k}_0$ be the circle with radius r and the point a as center. Then, reflection in $\mathfrak{k}_0$ is understood to be the transition from a point $z \neq a$ in the plane, to that point z' which is on the same ray issuing from a as z is, and is such that the product of the distances $|z - a|$ and $|z' - a|$ is equal to r^2. The reflection of a, however, is again to be regarded as the point ∞. If $\mathfrak{k}_0$ is a straight line, reflection in it shall have the elementary meaning. These somewhat more general reflections have, of course, properties entirely analogous to those possessed by reflection in the unit circle. We prove the following theorem concerning them:

THEOREM. *Let a circle $\mathfrak{k}$ and two points z_1 and z_2 symmetric to it be reflected in a circle $\mathfrak{k}_0$, yielding the circle $\mathfrak{k}'$ and the points z_1' and z_2' , say. Then z_1' and z_2' are symmetric with respect to $\mathfrak{k}'$.*

For, any two circles which pass through the points z_1 and z_2 are (by 4f) orthogonal to $\mathfrak{k}$. But then their images $\mathfrak{k}_1'$ and $\mathfrak{k}_2'$ are orthogonal to $\mathfrak{k}'$, because of the circularity and isogonality. They therefore intersect (again by 4f) in two points symmetric with respect to $\mathfrak{k}'$.

18. Mapping by means of arbitrary linear functions

Let there be given, finally, an arbitrary linear function

$$(1) \qquad w = \frac{az + b}{cz + d}.$$

Then c and d must not both vanish. If $c = 0$, and hence $d \neq 0$, we have the entire linear function $w = (a/d)z + (b/d)$, whose mapping we are already familiar with. If $c \neq 0$, we can write (1) in the form

$$(2) \qquad w = -\frac{ad - bc}{c} \cdot \frac{1}{cz + d} + \frac{a}{c}.$$

From this we infer that (1) is identically constant (and hence, the mapping degenerates) if, and only if, the determinant of the four coefficients is equal to zero.[33] *We therefore assume that*

[33] This is obviously also true for the case $c = 0$.

(3) $$ad - bc \neq 0$$

for all linear functions of the form (1) *appearing in what follows.*
Then the mapping furnished by (1) can be obtained in three
steps:

1) by the mapping $z' = cz + d$,
2) by the mapping $z'' = 1/z'$, and
3) by the mapping $w = a_1 z'' + b_1$, with

$$a_1 = -\frac{ad - bc}{c}, \qquad b_1 = \frac{a}{c}.$$

The first and third are similarity mappings, the second is the
one investigated in §17. We therefore have immediately the
following *principal theorem*:

THEOREM 1. (1) *furnishes a one-to-one mapping of the full z-
sphere on the full w-sphere. This mapping is isogonal and circular.*

In particular, the point $z = -d/c$ goes over into $w = \infty$ (for,
it yields first $z' = 0$, then $z'' = \infty$, and, consequently, also
$w = \infty$), and $z = \infty$ goes over into $w = a/c$. It is therefore
reasonable to *stipulate*, as a supplement to §17(5), *that, when
considering linear functions,*

(4) $$\frac{a \cdot \infty + b}{c \cdot \infty + d} = \frac{a}{c}.$$

The point z, whose image is *preassigned* to be the point w, is
given, according to (1), by

(5) $$z = \frac{-dw + b}{cw - a}.$$

The linear function (5) is therefore called the *inverse* of (1).
The determinant of the coefficients of (5) is the same as that
of (1).

The angle between two curves at ∞ is understood to be, of
course, the angle at which they intersect *on the sphere* at the
north pole. The meaning of isogonality is also clear, then, if
the image point, or its original, or both, lie at ∞. Thus, *when
considering linear functions, the point ∞ is in no way singled out
from the other points to play an exceptional role.*

On the basis of the last theorem in §17, we can state, finally, the following one:

THEOREM 2. *Under every mapping of the form* (1), *the figure of a circle*[34] *and two points symmetric with respect to it goes over into the same kind of figure.*

For, the similarity transformations 1) and 3) certainly possess this property; and 2) likewise, because it is equivalent to the successive performance of two reflections, each of which, according to the Theorem of §17, possesses the property.

[34]The straight lines are to be included here.

CHAPTER V

NORMAL FORMS AND PARTICULAR LINEAR TRANSFORMATIONS

19. The group-property of linear transformations

Let us go from a z-plane to a $\mathfrak{z}$-plane by means of a first linear mapping

$$(1) \qquad \mathfrak{z} = \frac{a_1 z + b_1}{c_1 z + d_1} = l_1(z),$$

and thence to a w-plane by means of the linear mapping

$$(2) \qquad w = \frac{a_2 \mathfrak{z} + b_2}{c_2 \mathfrak{z} + d_2} = l_2(\mathfrak{z}).$$

Then a simple calculation shows that the direct transition from the z- to the w-plane is effected by the function

$$(3) \qquad w = \frac{az + b}{cz + d} = l(z),$$

whose four coefficients can be read off from the *"matrix equation"*

$$\begin{pmatrix} a & b \\ c & d \end{pmatrix} = \begin{pmatrix} a_2 & b_2 \\ c_2 & d_2 \end{pmatrix}\begin{pmatrix} a_1 & b_1 \\ c_1 & d_1 \end{pmatrix} = \begin{pmatrix} a_2a_1 + b_2c_1 & a_2b_1 + b_2d_1 \\ c_2a_1 + d_2c_1 & c_2b_1 + d_2d_1 \end{pmatrix}.^{35}$$

By compounding two linear mappings $\mathfrak{z} = l_1(z)$, $w = l_2(\mathfrak{z})$ we thus again obtain a linear mapping

$$(4) \qquad l(z) = l_2(l_1(z)) = l_2 l_1(z).$$

If l_1 and l_2 do not degenerate, neither does l. For, according to the multiplication theorem for determinants, or by a simple calculation, we find that

---

[35] The *rows* of the first matrix are "combined" with the *columns* of the next; i.e., the sum of the products of the corresponding elements of the two is formed,—just as when multiplying determinants.

53

$$\begin{vmatrix} a & b \\ c & d \end{vmatrix} = \begin{vmatrix} a_2 & b_2 \\ c_2 & d_2 \end{vmatrix} \cdot \begin{vmatrix} a_1 & b_1 \\ c_1 & d_1 \end{vmatrix},$$

and since neither factor is zero, the product is not zero. It is also easy to verify that this compounding or "symbolic multiplication" of linear functions is associative, i.e.,

$$(5) \qquad\qquad l_3(l_2l_1) = (l_3l_2)l_1 .$$

Every function has also an inverse; for, according to §18, (5), the inverse of (3) is the function

$$w = \frac{-dz + b}{cz - a}.$$

It is denoted by $l^{-1}(z)$. When compounded with $l(z)$, it yields the identity:

$$ll^{-1}(z) = l^{-1}l(z) \equiv z,$$

which corresponds to the coefficient array

$$\begin{pmatrix} 1 & 0 \\ 0 & 1 \end{pmatrix}^{36}.$$

On the basis of these facts, we can state the following

THEOREM. *The linear mappings form a group, if the compounding of linear functions is employed as group multiplication. The identity is the identity element of the group, inverse functions are inverse elements.*

20. Fixed points and normal forms

In §17 we already spoke of *fixed points* of a mapping. A fixed point is understood to be one which coincides with its image. If this is to be the case for a point z under the mapping

$$(1) \qquad\qquad w = \frac{az + b}{cz + d} = l(z),$$

z must satisfy

[36]Or $\begin{pmatrix} a & 0 \\ 0 & a \end{pmatrix}$, with $a \neq 0$.

(2) $\qquad \dfrac{az + b}{cz + d} = z \qquad$ or $\qquad cz^2 - (a - d)z - b = 0.$

This is a quadratic equation in z, whose coefficients all vanish only if the mapping is the identity ($a = d \neq 0$, $b = c = 0$). We therefore have immediately

THEOREM 1. *A linear mapping which is not the identity has at most two fixed points.*—Hence, *if a linear mapping is known to have at least three fixed points, it must be the identity.*

If $c \neq 0$, so that the mapping is a *fractional* linear one, both fixed points (which, of course, may also coincide) are finite. If $c = 0$, in which case we have an *entire* linear mapping, at least one of the fixed points lies at ∞ (this follows already from §16). If, moreover, $a = d$ (but $b \neq 0$), we are dealing with a translation, which leaves only the point ∞ fixed. Hence, as a supplement to the preceding theorem, we have

THEOREM 2. *The point ∞ is a fixed point if, and only if, the linear mapping is entire; it is the only fixed point if, and only if, the mapping is a translation.*

Through the use of the fixed points, one can acquire an even more vivid insight into the nature of the linear mapping.

1) First, let

(3) $\qquad\qquad\qquad\qquad w = az + b$

be an entire linear mapping which is not a translation (consequently $a \neq 1$). In addition to the fixed point ∞, it has the finite fixed point

(4) $\qquad\qquad\qquad\qquad \zeta = \dfrac{b}{1 - a}.$

By using ζ, the mapping (3) can be brought into the form

(5) $\qquad\qquad\qquad w - \zeta = a(z - \zeta),$

from which it can be interpreted as follows: In the z-plane, first perform the translation $z - \zeta$ (it moves the point ζ to the origin). Now effect the rotary stretching (a), and finally bring the point 0 back to ζ by a translation. Thus we see that the mapping (3) signifies simply a rotary stretching (a) with the fixed point ζ as center! The mapping has become perfectly clear through this interpretation. In particular, it is evident that the pencil of straight lines through ζ goes over into itself as a whole, and the same holds for the family of concentric circles about ζ as center: under a pure rotation ($|a| = 1$), each of the circles individually; under a pure stretching ($a > 0$), each of the straight lines individually.

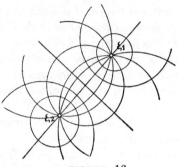

FIGURE 16

2) Now let $c \neq 0$, so that both fixed points, call them ζ_1 and ζ_2, are finite. Moreover, let $\zeta_1 \neq \zeta_2$. Then, it follows immediately from the circularity of the mapping, that the pencil of circles through ζ_1 and ζ_2 goes over into itself as a whole (Fig. 16). More particular information about this can be obtained as follows:

By the mapping

(6)
$$\mathfrak{z} = \frac{z - \zeta_1}{z - \zeta_2} = l_0(z),$$

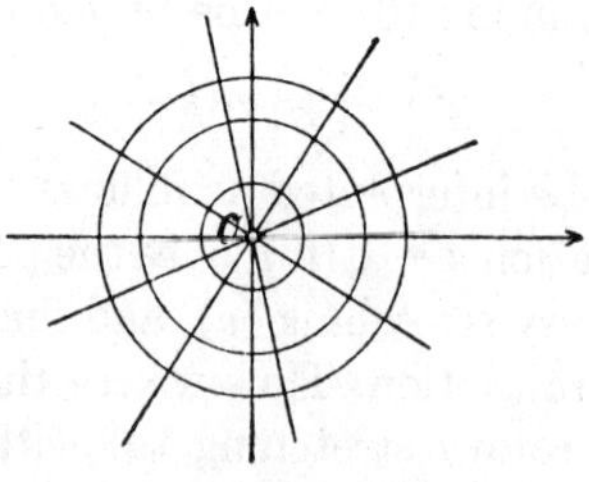

FIGURE 17

which brings ζ_1 to 0 and ζ_2 to ∞, the pencil in Fig. 16 is mapped into the pencil of straight lines in Fig. 17: the circles through 0 and ∞. Let the first lie in the z-plane, the second, in the $\mathfrak{z}$-plane. If we imagine the first to lie in the w-plane, the second, in a $\mathfrak{w}$-plane, then, analogously,

$$(7) \qquad \mathfrak{w} = \frac{w - \zeta_1}{w - \zeta_2} = l_0(w)$$

maps Fig. 16 into Fig. 17. Because of the group-property of the linear functions, a linear mapping of the $\mathfrak{z}$-plane on the $\mathfrak{w}$-plane, namely, the mapping

$$(8) \qquad \mathfrak{w} = l_0 l l_0^{-1}(\mathfrak{z}),$$

is effected hereby and by the mapping (1). Since (8) has 0 and ∞ as fixed points, it is, according to 1), a rotary stretching with 0 as center, and therefore has the simple form $\mathfrak{w} = \mathfrak{a}\mathfrak{z}$, where $\mathfrak{a}$ is a certain complex number. Consequently, by using the fixed points, (1) can be brought into the following *normal form*:

$$(9) \qquad \frac{w - \zeta_1}{w - \zeta_2} = \mathfrak{a}\frac{z - \zeta_1}{z - \zeta_2}.$$

The value of $\mathfrak{a}$ is found immediately by setting $z = \infty$, which yields $w = a/c$. Hence,

$$(10) \qquad \mathfrak{a} = \frac{a - c\zeta_1}{a - c\zeta_2}.$$

The argument carried out shows more at the same time: Since there is a family of circles orthogonal to the pencil of straight lines in Fig. 17, namely, the family of circles about 0 as center, there also exists a family of circles orthogonal to the pencil of "circles *through* ζ_1 and ζ_2" in Fig. 16. This family we shall call, for brevity, the family of "circles *about* ζ_1 and ζ_2." It, too, and hence the complete Fig. 16, goes over into itself as a whole, under the mapping (1). Beyond this we can say more precisely:

a) If $|\mathfrak{a}| = 1$, then $\mathfrak{w} = \mathfrak{a}\mathfrak{z}$ is a pure rotation, and, consequently, each of the circles of the second family goes over into itself as a whole, while those of the first family are permuted among themselves. The mapping (1) is then said to be *elliptic*.

b) If a is real and positive, it is just the reverse. The mapping (1) is then said to be *hyperbolic*.

c) If $|a| \gtrless 1$ and a is not a positive real number, the mapping is said to be *loxodromic*. It is obtained by carrying out steps a) and b) in succession.[37]

3) Let $c \neq 0$ once more, but now suppose that $\zeta_2 = \zeta_1$. The mapping is then said to *parabolic*.[38] The totality of circles through the fixed point—call it ζ—goes over into itself as a whole. A family of circles through ζ, which have there a com-

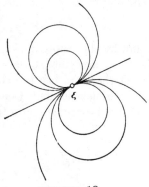

FIGURE 18

mon tangent (see Fig. 18), likewise remains unchanged as a whole. This, as well as further details, are again seen more clearly if the fixed point (in the z-plane *and* in the w-plane) is sent to ∞ by the auxiliary mappings

(11) $$ \mathfrak{z} = \frac{1}{z - \zeta}, \qquad \mathfrak{w} = \frac{1}{w - \zeta}, $$

respectively. The circles in question then go over into the straight lines of the plane; those circles having a common tan-

[37]If the auxiliary plane of Fig. 17 is used, the individual steps can be followed even more closely.

[38]The same terminology, of course, is used for *entire* linear mappings: a translation is said to be *parabolic*; a rotary stretching with the fixed point as center (cf. (5)) is said to be *loxodromic*; a pure stretching, *hyperbolic*; a pure rotation, *elliptic*.

gent, into a family of parallels. Again $\mathfrak{z}$-plane and $\mathfrak{w}$-plane are mapped linearly on each other. But since ∞ is the only fixed point under this mapping, the two planes arise from each other by means of a *translation* $\mathfrak{w} = \mathfrak{z} + \mathfrak{b}$. Hence, in the parabolic case, (1) can be brought into the form

$$(12) \qquad \frac{1}{w - \zeta} = \frac{1}{z - \zeta} + \mathfrak{b};$$

and since $z = \infty$ and $w = a/c$ correspond, we must have

$$(13) \qquad \mathfrak{b} = \frac{c}{a - c\zeta}.$$

The family of circles lying in the z-plane, passing through ζ, and having there a common tangent, is transformed by (11) into a family of parallels in the $\mathfrak{z}$-plane. This last family is left unchanged, as a whole, by the translation ($\mathfrak{b}$), and hence, returns, by virtue of (11), to the initial family.

21. Particular linear mappings. Cross ratios

Theorem 1 of §18, in particular, the fact that circles are always transformed into circles by linear mappings, remains the most important of all that precedes. We shall now investigate more closely how this takes place. To this end we first prove

THEOREM 1. *Three given distinct points z_1 , z_2 , z_3 can always be carried into three prescribed distinct points w_1 , w_2 , w_3 by one, and only one, linear mapping, $w = l(z)$.*[39]

Proof: The equation

$$(1) \qquad \frac{w - w_1}{w - w_3} : \frac{w_2 - w_1}{w_2 - w_3} = \frac{z - z_1}{z - z_3} : \frac{z_2 - z_1}{z_2 - z_3}$$

defines a definite linear function $w = l(z)$. For, on the left we have a linear function of w, on the right, a linear function of z. If we call these $l_1(w)$ and $l_2(z)$, respectively, then $w = l(z) = l_1^{-1} l_2(z)$. Here we must agree, in accordance with §18 (4), that if one of the points z_ν or w_ν is the point ∞, the quotient of those two differences which contain this point is to be replaced by 1.

[39]One of the points z_ν , as well as one of the w_ν , ($\nu = 1, 2, 3$), may also be the point ∞.

This function $w = l(z)$, now, accomplishes the desired end. For, $l_2(z)$ assumes, for $z = z_1$, z_2, z_3, the respective values 0, 1, ∞, and $l_1(w)$ takes on these values for $w = w_1$, w_2, w_3, respectively. Hence, $l(z_\nu) = w_\nu$, ($\nu = 1, 2, 3$). If the linear function $w = L(z)$ accomplishes the same, then the linear function $L^{-1}l(z)$ obviously has the three distinct fixed points z_1, z_2, z_3, and, consequently, according to §20, Theorem 1, is the identity. Therefore $L(z) = l(z)$. This completes the proof.

Now, an oriented circle (including the oriented straight line) is uniquely determined by three (distinct) points given in a definite order. Hence, from Theorem 1 follows immediately the further

THEOREM 2. *A given oriented circumference of a circle in the z-plane or on the z-sphere can always be mapped by one, and only one, linear function, into a given oriented circumference of a circle in the w-plane, in such a manner, that three given points of the z-circle thereby go over into three given points of the w-circle, provided that on both circles the points succeed one another in the sense of the orientation.*

The complex plane is divided by a circle (or a straight line) into two parts. That one of them which lies *to the left* of the orientation will be called *"the interior"* of the circle, and the other one, *"the exterior"* of the circle.[40] The complex sphere is divided by a circle into two spherical caps. That one which, viewed inside the sphere, lies *to the left* of the orientation we shall call *the interior* of the circle, the other, *the exterior* of the circle, so that interiors of circles correspond to interiors of circles under stereographic projection. Since the mapping of full spheres by means of linear functions is one-to-one and, moreover, isogonal without reversion of angles, there follows, now, as a supplement to Theorem 2:

THEOREM 3. *The linear function mentioned in Theorem 2 maps the interior of the z-circle in a one-to-one manner on the interior of the w-circle; and likewise, naturally, the exterior of the first on the exterior of the second.*

To express briefly the hereby established property of linear

[40]Thus, if, e.g., the axis of imaginaries is oriented from bottom to top, the left half-plane is the interior, the right half-plane is the exterior.

functions, that if they map two oriented circles into one another they also map their interior regions (and likewise their exterior regions) on one another in a one-to-one manner, we say that *regions are preserved* under linear mappings.

Examples of these, and the mappings mentioned later on, follow in §22.

The peculiar expressions appearing in (1) are called *cross ratios*. The following is a more precise

DEFINITION. *Let z_1 , z_2 , z_3 , z_4 be four distinct points on the sphere. Then the expression*

$$(2) \qquad (z_1 , z_2 ; z_3 , z_4) = \frac{z_4 - z_1}{z_4 - z_3} : \frac{z_2 - z_1}{z_2 - z_3}$$

shall be called their cross ratio. If one of the points lies at ∞ , then the agreement made above comes into force.[41]

From the proof of Theorem 1 now follows immediately

THEOREM 4. *The cross ratio of four points remains invariant under linear mappings.*

That is to say: If the four points z_ν go over into the respective points w_ν , ($\nu = 1, 2, 3, 4$) under the mapping $w = l(z)$, then

$$(w_1 , w_2 ; w_3 , w_4) = (z_1 , z_2 ; z_3 , z_4).$$

For, since $w = l(z)$ performs what is required in Theorem 1, it must be the function given by (1). It also carries z_4 into w_4 . Hence, for $z = z_4$, $w = w_4$, (1) immediately yields the assertion.

An oriented circle can be given by means of *one* point of the circumference and a pair of points symmetric with respect to the circle, instead of by means of three points of the circumference. This yields, in connection with Theorem 1,

THEOREM 5. *An oriented z-circle can always be linearly*[42]

[41]The order in which the four points are taken is not essential, but, of course, once it has been chosen, it must be retained. If the four points are permuted in all possible ways, we do not obtain 24 distinct values of the cross ratio, but, on the contrary, at most 6. If one of the values is equal to δ, the others are $1/\delta$, $1 - \delta$, $1/(1 - \delta)$, $\delta/(\delta - 1)$, and $(\delta - 1)/\delta$. These values may coincide in part.

[42]It can be shown that *the most general function which maps the interior of one circle in a one-to-one and conformal manner on the interior of another circle, is a linear function.* See, e.g., L. R. Ford, *Automorphic Functions,* New York, 1929, p. 32.

mapped, in one, and only one, way, into an oriented w-circle in such a manner, that a given boundary point z_1 and a given interior point z_0 of the z-circle thereby go over into correspondingly situated, prescribed points w_1 and w_0.

Proof: Let z_0', w_0' be the reflections of z_0, w_0 in their respective circles. Then, according to §18, Theorem 2, a linear function which accomplishes what is required must also carry z_0' into w_0'. Hence, only the linear function resulting from

$$(3) \qquad (w_1 , w_0 ; w_0' , w) = (z_1 , z_0 ; z_0' , z)$$

can accomplish the desired end, and, according to the preliminary remark, this is also the case.

22. Further examples

1. *Mapping the upper half-plane (UH) on the unit circle (UC).*

a) If we require, say, that the interior point i of the *UH* go over into the center of the *UC*, and that the boundary point 0 of the *UH* go over into the boundary point -1 of the *UC*, then, according to §21, Theorem 5, the mapping is uniquely determined. Since it sends $-i$ to ∞, the points $z = i, 0, -i$ go over into $w = 0, 1, \infty$, respectively.[43] Hence, according to §21, Theorem 1,

$$(1) \qquad \frac{w - 0}{-1 - 0} = \frac{z - i}{z + i} : \frac{0 - i}{0 + i} \qquad \text{or} \qquad w = \frac{z - i}{z + i}$$

is the required mapping. For every real z, $|w| = 1$, as is easily verified. Conversely, by means of the inverse function

$$(2) \qquad z = -i \frac{w + 1}{w - 1} ,$$

the *UC* of the *w*-plane is mapped on the upper *z*-half-plane. The further details of the mapping become more vivid if we ask, which curves in the *UH* go over into the radii of the *UC*, and which curves go over into the circles which have the same

---

[43]We purposely order the points in such a manner, that $w_3 = \infty$. For then $(w_1 , w_2 ; w_3 , w)$ has the simple form $(w - w_1)/(w_2 - w_1)$, and thus contains the variable w, for which we must finally solve, only in the numerator.

center as the UC but are smaller than the latter. Because of the circularity of the mapping, we immediately obtain: Referring to the circles in the z-plane which pass through $+i$ and $-i$, those parts of them which lie in the UH are transformed into the radii of the UC; and the circles "about $+i$ and $-i$" which are orthogonal to the first pencil of circles and lie in the UH are transformed into the circles which have the same center as the UC but are smaller than the latter.

By means of this mapping, moreover, the first quadrant of the z-plane is mapped on the lower half of the UC; thus, a quarter-plane, on a semicircle.[44]

b) If we require, somewhat more generally, that the point z_0, $(\Im(z_0) > 0)$, in the UH go over into the origin, and that the boundary point α, (α real), go over into the boundary point -1, then

$$(3) \qquad w = -\frac{\alpha - \bar{z}_0}{\alpha - z_0} \cdot \frac{z - z_0}{z - \bar{z}_0} = c\,\frac{z - z_0}{z - \bar{z}_0}, \qquad (|\,c\,| = 1),$$

accomplishes what is required. The further details are entirely similar to those under a).

c) The requirement that the three boundary points 0, 1, ∞ of the UH go over into the boundary points i, -1, $-i$ of the UC, also determines the mapping uniquely. We find (and it can be verified subsequently by substituting the z-values):

$$(4) \qquad w = -i\,\frac{z - i}{z + i}.$$

2. *The UH of the z-plane is to be mapped on the UH of the w-plane in such a manner, that the points $z = \infty$, 0, 1 go over into the points $w = 0$, 1, ∞, respectively.* The mapping is hereby uniquely determined. Proceeding as in 1., we find

$$(5) \qquad w = -\frac{1}{z - 1}.$$

[44]It is earnestly recommended that the reader make simple sketches for all of the mappings discussed, letting corresponding points and parts of boundaries or regions become clear by using the same colors or hatchings.

Under this mapping, what becomes of the parallels to the axis of reals; the parallels to the axis of imaginaries; the two quadrants of the half-plane? What are the answers to the "inverse" questions?

3. *The exterior of the UC is to be mapped on the right half-plane.* If we send, let us say, the points

$$z = 1, -i, -1 \quad \text{to} \quad w = i, 0, \infty,$$

respectively, then in each of the planes the region in question lies to the left of the orientation given in this manner. Hence, the mapping

$$(6) \qquad w = i + \frac{z - 1}{z + 1} = \frac{(1 + i)z + (-1 + i)}{z + 1}$$

does what is required. We leave it to the reader to determine what becomes of the circles which have the same center as the *UC* but are larger than the latter, and what becomes of those parts of the rays issuing from 0, which lie outside the *UC*.

4. *The UC is to be mapped on itself* in such a manner, that the interior point z_0, ($| z_0 | < 1$), is transformed into the center.

If z_0 is to go over into 0, the reflection of z_0 in the unit circle, i.e., the point $z_0' = 1/\bar{z}_0$, must be sent to ∞. Hence, the linear function we are looking for must have the form

$$w = c\, \frac{z - z_0}{z - (1/\bar{z}_0)} \quad \text{or} \quad w = c'\, \frac{z - z_0}{\bar{z}_0 z - 1}.$$

Now, the radius of the image circle will again be equal to 1 if, and only if, the image of the point $+1$ has the absolute value 1:

$$\left| c'\, \frac{1 - z_0}{\bar{z}_0 - 1} \right| = | c' | = 1.$$

Hence, in particular, the function

$$w = \frac{z - z_0}{\bar{z}_0 z - 1}$$

yields the required mapping.

5. *Two circles which have no point in common can always be transformed, by a linear mapping, into two concentric circles.* For,

two circles of the first kind (whether one encloses the other or not) can always be regarded (in precisely one way) as two circles of a pencil of the kind described in connection with Fig. 16; i.e., there is precisely one pair of points ζ_1 and ζ_2 such that the given circles belong to the pencil of "circles about ζ_1 and ζ_2 ."[45] The mapping

$$w = \frac{z - \zeta_1}{z - \zeta_2}$$

then obviously performs what is required.

Finally, we prove the following *Theorem*:

6. *The cross ratio of four points is real if, and only if, the points lie on a circle (or a straight line).* For if the cross ratio is to be real,

$$\text{am } \frac{z_4 - z_1}{z_4 - z_3} \quad \text{and} \quad \text{am } \frac{z_2 - z_1}{z_2 - z_3}$$

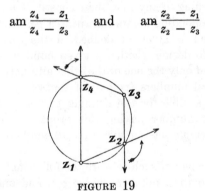

FIGURE 19

must either be equal or differ only by $\pm\pi$. The first amplitude signifies the angle through which the direction extending from z_3 to z_4 must be rotated in the positive sense until it coincides with the direction leading from z_1 to z_4 , and the second amplitude has an analogous meaning. Elementary theorems on peripheral angles now establish the validity of the theorem (see Fig. 19),—irrespective of whether the pair of points z_2 , z_4 are separated by the pair z_1 , z_3 or not.

[45]The "base points" ζ_1 and ζ_2 of the pencil are found by drawing the line of centers of the circles. ζ_1 and ζ_2 then separate harmonically the pair of points of intersection of each of the circles with this line.

CHAPTER VI

POINT SETS AND SETS OF NUMBERS

23. Point sets

If a finite or an infinite number of complex numbers are selected according to any rule, these constitute a *set of numbers*, and the corresponding points constitute a *point set*.[46] Such a set $\mathfrak{M}$ is regarded as given or defined, if the rule for selecting enables one to decide whether a given number belongs to the set or not (and only the one or the other alternative is possible). The individual numbers (points) of the set are called its *elements*. It is possible for the defining property of a set to be such that no number having this property exists—we then speak of the *empty* set—or such that *all* numbers belong to the set.

Simple *examples* of such sets are the following:

$\mathfrak{M}_1$. All complex numbers whose real and imaginary parts are integers. The points of this set are called the *lattice points* of the plane.

$\mathfrak{M}_2$. All complex numbers whose real and imaginary parts are rational.

$\mathfrak{M}_3$. All real numbers.

$\mathfrak{M}_4$. All numbers of the form $1 \pm (1/n)$, where n is a natural number.

$\mathfrak{M}_5$. All numbers of the form $(1/m) + (i/n)$, where m and n are natural numbers.

$\mathfrak{M}_6$. All complex numbers z for which $|z| < 1$.

[46]In this chapter, we use only the *plane* for graphical representation. The point ∞ will not be employed for a while.

Each of the relations considered in §13, 4.–8. defines a certain point set.

A set $\mathfrak{M}$ is said to be *bounded*, if there exists a positive number K such that

$$| z | \leqq K$$

"for all z of the set" (i.e., *for every* number z which belongs to $\mathfrak{M}$). Such a number K is then called a *bound* for the (moduli of the numbers of the) set. In the contrary case, $\mathfrak{M}$ is said to be unbounded. Of the sets just given as examples, $\mathfrak{M}_4$, $\mathfrak{M}_5$, and $\mathfrak{M}_6$ are bounded, the others are not. The totality of points which do *not* belong to $\mathfrak{M}$ constitute the *complementary set* or *complement* of $\mathfrak{M}$.

If a point ζ of the plane has the property that an infinite number of points of $\mathfrak{M}$ lie in *every* ϵ-neighborhood (see §13,6.) of ζ, the latter is called a *limit point* of $\mathfrak{M}$. $\mathfrak{M}_1$ has no limit point, whereas *every* point of the plane is a limit point of $\mathfrak{M}_2$. The point $+1$ is the only limit point of $\mathfrak{M}_4$; the limit points of $\mathfrak{M}_5$ are 0 and all points of the form $1/m$ and i/n (m, n are natural numbers). Every point ζ with $| \zeta | \leqq 1$ is a limit point of $\mathfrak{M}_6$.

In §§24 and 25 we shall prove the important *Bolzano-Weierstrass theorem*:

THEOREM. *Every bounded infinite (i.e., consisting of an infinite number of points) point set has at least one limit point.*

A point belonging to $\mathfrak{M}$ is said to be *"isolated,"* if there exists an ϵ-neighborhood of the point containing no other point of $\mathfrak{M}$. It is called an *interior point* of $\mathfrak{M}$, if an ϵ-neighborhood of the point belongs entirely to $\mathfrak{M}$. $\mathfrak{M}_1$, $\mathfrak{M}_4$, and $\mathfrak{M}_5$ consist wholly of isolated points, $\mathfrak{M}_6$ contains only interior points.

A point ζ of the plane (ζ may or may not belong to $\mathfrak{M}$) is called a *boundary point* of $\mathfrak{M}$, if there is at least one point which belongs to $\mathfrak{M}$ and at least one which does not belong to $\mathfrak{M}$ in every ϵ-neighborhood of ζ. All the sets given as examples above, except $\mathfrak{M}_6$, consist exclusively of boundary points. Those, and only those, points ζ with $| \zeta | = 1$ are boundary points of $\mathfrak{M}_6$.

A set is said to be *closed*, if it contains all its limit points. A set is said to be *open*, if every one of its points is an interior point of the set. $\mathfrak{M}_1$ and $\mathfrak{M}_3$ are closed, the rest of the above sets are not; $\mathfrak{M}_6$ is open.

24. Sets of real numbers

If, in the considerations of the preceding paragraph, we re-strict ourselves to the totality of real numbers, we arrive at the corresponding concept of *sets of real numbers* or *real point sets*. The definitions remain essentially the same. It is to be noted, merely, that the complement of $\mathfrak{M}$ is only the set of all *real* numbers which do not belong to $\mathfrak{M}$, and that an ϵ-neigh-borhood of a real number ξ consists of only those *real* numbers x for which $|x - \xi| < \epsilon$. Except for these, all definitions re-main exactly the same.[47] Nevertheless, several new details arise, due to the fact that the real numbers form an *ordered* set:

A real set is said to be *bounded on the left*, if a number[48] K_l exists such that $x \geqq K_l$ for all x of the set. If all $x \leqq K_r$, the set is said to be *bounded on the right*. K_l is called a *lower bound*; K_r, an *upper bound*. The former may be replaced by any smaller number, the latter, by any larger number, but not conversely. Of all the lower bounds, however, there is a greatest; i.e., there exists a number γ with the following two properties:

1. No point of the set lies to the left of γ; briefly: there is

$$no \ x < \gamma.$$

2. *At least one* point of the set lies to the left of every greater number; in other words: for every $\epsilon > 0$, there is

$$at \ least \ one \ x < \gamma + \epsilon.$$

This number γ is called the *greatest lower bound* of the set (abbreviated: g.l.b.). We prove

THEOREM 1. *Every set which is bounded on the left possesses a well-determined greatest lower bound* γ.

Proof: Divide the totality of all real numbers into two classes $\mathfrak{A}$, $\mathfrak{A}'$. Into the class $\mathfrak{A}$, put all real numbers a for which *no*

---

[47]It is to be observed, however, that every real set *may* also be regarded as a set of complex numbers, since the real numbers are contained in the system of complex numbers.

[48]The "numbers" in this paragraph shall always be real numbers.

$x < a$; into the class $\mathfrak{A}'$, put every number a' for which *at least one* $x < a'$. By hypothesis, $\mathfrak{A}$ and $\mathfrak{A}'$ are not empty, and we have always $a < a'$, since otherwise there would exist an $x < a$.

Let γ be the real number defined by this cut. Then γ is the greatest lower bound of $\mathfrak{M}$. For if there were an $x < \gamma$, x would also be less than every number between x and γ, which number, being less than γ, would belong to $\mathfrak{A}$. Thus, there would be an $x < a$, which is impossible. Hence, there is *no* $x < \gamma$. On the other hand, if $\epsilon > 0$, then $\gamma + \epsilon$ belongs to $\mathfrak{A}'$, and, consequently, there exists *at least one* $x < \gamma + \epsilon$, Q.E.D.

Similarly, the *least upper bound* (l.u.b.) of a set is defined as the number γ' with the following two properties:

1. There is *no* $x > \gamma'$.
2. However $\epsilon > 0$ be chosen, there is *at least one* $x > \gamma' - \epsilon$.

Concerning the least upper bound, we have

THEOREM 2. *Every set which is bounded on the right possesses a definite least upper bound* γ'.

A set which is bounded on both sides thus possesses a definite greatest lower bound and a definite least upper bound. The two numbers themselves need not be points of the set. If a set is not bounded on the left, we say also that its greatest lower bound is $-\infty$; if it is not bounded on the right, we say that its least upper bound is $+\infty$.

The *Bolzano-Weierstrass* theorem, which was already mentioned in §23, can now be proved in the real domain. The *proof* is quite similar to the one just given:

We again divide the totality of all real numbers into two classes $\mathfrak{A}$, $\mathfrak{A}'$. Into the class $\mathfrak{A}$, put every number a having the property that no points, or at most a finite number of points, of the set lie to the left of a:

at most finitely many $x < a$.

Into the class $\mathfrak{A}'$, put every number a' having the property that an infinite number of points of the set lie to the left of a':

infinitely many $x < a'$.

It is immediately evident that this classification represents a

Dedekind cut. Let it define the number μ. Then $\mu - \epsilon$ belongs to $\mathfrak{A}$. Consequently, at most a finite number of points of the set lie to the left of $\mu - \epsilon$. On the other hand, $\mu + \epsilon$ belongs to $\mathfrak{A}'$, and therefore there are infinitely many x of the set which are less than $\mu + \epsilon$. Hence, infinitely many x of the set must lie between $\mu - \epsilon$ and $\mu + \epsilon$; i.e., μ is a limit point of the set.

Since at most a finite number of points of the set lie to the left of $\mu - \epsilon$, there is certainly no further limit point there; i.e., μ is the *least* limit point, or the one *farthest to the left*, and is therefore designated as the *lower limit* or *limes inferior* (abbreviated: $\underline{\lim}$ or lim inf).

It can be proved in an entirely similar manner, that there exists a greatest limit point, or one lying farthest to the right, μ', which is called the *upper limit* or *limes superior* ($\overline{\lim}$, lim sup). It is characterized by the following two properties:

1) There are at most a finite number of $x > \mu' + \epsilon$.

2) There are an infinite number of $x > \mu' - \epsilon$, no matter how the positive number ϵ may be chosen.

It is evident that always $\mu \leqq \mu'$. These points need not belong to the set. Together they are called the *principal limits* of the set.

If a set is not bounded on the left, then we designate $- \infty$ as its lower limit; likewise, $+ \infty$ as its upper limit, if it is not bounded on the right. Finally, if the set is bounded on the right but not on the left, and if it has *no* finite limit point *whatsoever*, it is reasonable to call $- \infty$ its lim sup, and, in the "mirror image" of this case, $+ \infty$ its lim inf.

The set of real numbers which lie between two real numbers a and b, $(a < b)$,—they fill out a segment of the number axis,— is termed the *interval $a \ldots b$*. It is called *closed* or *open*, according as the end points are regarded as belonging to the set or not. The first is denoted by $\langle a, b \rangle$, the second, by (a, b).

25. *The Bolzano-Weierstrass theorem*

Leaning on the *Bolzano-Weierstrass theorem* in the real domain (§23), we can now also prove it in the complex domain. Let $\mathfrak{M}$ be a bounded infinite point set in the z-plane. Then we can demonstrate the existence of a limit point ζ of this set as follows:

The set of real numbers $\Re(z)$, (z in $\mathfrak{M}$), is *either* again a bounded infinite (real) set, and then has, according to §24, at least one limit point ξ; *or* it is finite. In the latter case, however, there must be among its finitely many elements at least one, call it ξ, such that $\Re(z) = \xi$ for an infinite number of z of the set. In either case, then, there exists a real number ξ such that, for *every* $\epsilon > 0$,

$$\xi - \epsilon < \Re(z) < \xi + \epsilon$$

for an infinite number of z of the set. We say briefly: In every *ϵ-strip* about ξ, there are an infinite number of z of the set. Now we execute a Dedekind cut for the *ordinates*. We divide all real numbers into two classes $\mathfrak{B}$, $\mathfrak{B}'$. Into $\mathfrak{B}$ we put all real numbers b with the property: In *every* ϵ-strip about ξ, there are an infinite number of z of the set with $\Im(z) > b$. Into $\mathfrak{B}'$ we put all numbers b' for which this is not the case. This classification ($\mathfrak{B}|\mathfrak{B}'$) is obviously a cut. Let it define the real number η. *Then $\zeta = \xi + i\eta$ is a limit point of $\mathfrak{M}$.* For, if $\epsilon > 0$ is given arbitrarily, $\eta + \epsilon$ belongs to the class $\mathfrak{B}'$. There is therefore an ϵ'-strip about ξ, with $0 < \epsilon' < \epsilon$, such that in this strip *only a finite* number of z of the set have an imaginary part $\Im(z) > \eta + \epsilon$. Since, however, $\eta - \epsilon$ belongs to $\mathfrak{B}$, there are an *infinite number* of z in this strip with $\Im(z) > \eta - \epsilon$. Hence, an infinite number of z of the set lie in the rectangle

$$\xi - \epsilon' < \Re(z) < \xi + \epsilon', \qquad \eta - \epsilon < \Im(z) < \eta + \epsilon,$$

and, consequently, also an infinite number in the square ϵ-neighborhood of ζ, Q.E.D.

CHAPTER VII

SEQUENCES OF NUMBERS.
INFINITE SERIES

26. Sequences of complex numbers

If, by virtue of an unambiguous rule, there corresponds to every natural number 1, 2, 3, ... one definite complex number z_1, z_2, z_3, ..., respectively, there results a *sequence of numbers*, which is denoted for brevity by $\{z_n\}$ or $\{z_1, z_2, ...\}$. The z_n are called its *terms*. The values of the terms need not be distinct. Often a "zeroth" term is placed at the beginning of the sequence, as its leading term. Simple *examples* are the following:

1. $\{a^n\}$, i.e., the sequence of numbers 1, a, a^2, ..., a^n, ..., where a is a given number.

2. $\{1/n\}$, i.e., the sequence of numbers 1, 1/2, 1/3, ..., $1/n$, ...[49]

3. The sequence $\{z_n\}$, with $z_0 = 1$, $z_1 = i$, $z_n = \frac{1}{2}(z_{n-1} + z_{n-2})$ for $n \geqq 2$.

The points which correspond to the numbers z_n constitute a *sequence of points*. If one and the same point appears several times or an infinite number of times in the sequence, "it counts" several or an infinite number of times as a point of the sequence.

Conversely, if $\mathfrak{M}$ is an (infinite) point set, and if it is possible to designate the points as z_1, z_2, ... in such a manner, that *every* point of $\mathfrak{M}$ is thereby given a number, then $\mathfrak{M}$ is called an *enumerable* point set. The sets $\mathfrak{M}_1$, $\mathfrak{M}_2$, $\mathfrak{M}_4$, and $\mathfrak{M}_5$ in §23 are enumerable, $\mathfrak{M}_3$ and $\mathfrak{M}_6$ are not. (We shall not consider here a proof of this assertion.) In the case of a point set, we assume, of course, that every pair of its elements are distinct; this need not be true for the terms of an arbitrary sequence. We may, therefore, say also: A sequence of numbers is an

---

[49]Because of the form of the general term, it goes without saying, that the initial value of this sequence must be $n = 1$, not $n = 0$. The like is often to be noted in what follows.

enumerable point set (which has been enumerated in a definite manner), where one and the same point, however, is permitted to count several times or even infinitely often. The terminology introduced for point sets, consequently, carries over, with suitable interpretation, to sequences of numbers.

Thus, a number ζ is designated as a *limit point* of a sequence $\{z_n\}$ if, for every $\epsilon > 0$, an infinite number of z_n lie in the ϵ-neighborhood of ζ; in other words, if

$$(1) \qquad |z_n - \zeta| < \epsilon$$

for *infinitely many* n. It must now be noted, however, that a value ζ which appears infinitely often in the sequence is to be regarded as a limit point of the sequence. The *Bolzano-Weierstrass theorem* then says:

Every bounded sequence of numbers, $\{z_n\}$, has at least one limit point ζ.

The case in which it has *only one* limit point is of particular interest. The relation (1) then holds for *all sufficiently large values of n*. This is also expressed briefly by saying that (1) is valid for *nearly all n*, or for all n *after a certain one*, say for all $n > n_0 = n_0(\epsilon)$.[50] In this case, ζ is called the *limit* of the sequence $\{z_n\}$, and we write

$$\lim z_n = \zeta \qquad \text{or} \qquad z_n \to \zeta,$$

with or without the addition: "as $n \to +\infty$." The sequence itself is said to be *convergent* with the limiting value ζ, or to *tend to ζ*.

Cauchy's general convergence principle states a necessary and sufficient condition for the occurrence of this case:

THEOREM 1. *A necessary and sufficient condition for the sequence $z_0, z_1, z_2, \ldots$ to have a limit is that, for arbitrarily given $\epsilon > 0$, a number $n_0 = n_0(\epsilon)$ can be assigned, such that*

$$|z_{n+p} - z_n| < \epsilon$$

for all $n > n_0$ and all $p > 0$. (More briefly: *Nearly all z_n must have a distance of less than ϵ from each other.*)

---

[50] The last expression is used to indicate that the n_0th term, beyond which the relation (1) is valid, depends on the choice of ϵ.

Proof: 1. Suppose $z_n \to \zeta$. Then, $\epsilon > 0$ being given,

$$| z_n - \zeta | < \tfrac{1}{2}\epsilon$$

for $n > n_1(\epsilon)$. Hence, for all $n > n_1$ and all $p > 0$,

$$| z_{n+p} - z_n | = | (z_{n+p} - \zeta) - (z_n - \zeta) | < \epsilon,$$

—the last according to §11 (2). The condition is therefore necessary.

2. If, conversely, (2) is satisfied, the sequence $\{z_n\}$ is bounded. For, let us choose $\epsilon = 1$, say. Then, to this choice of ϵ there corresponds an n_1 such that, for all $n > n_1$,

$$| z_n - z_{n_1} | < 1, \quad \text{and hence,} \quad | z_n | < | z_{n_1} | + 1.$$

The greatest of the numbers $| z_1 |, | z_2 |, \ldots, | z_{n_1-1} |, | z_{n_1} | + 1$ is a bound for the set. Now, according to the Bolzano-Weierstrass theorem, $\{z_n\}$ has at least one limit point, ζ. If it has yet a second limit point $\zeta' \neq \zeta$, then $\epsilon = \tfrac{1}{3} | \zeta' - \zeta |$ would be a positive number; and an infinite number of z_n would lie in the ϵ-neighborhood of ζ, an infinite number of others, in the ϵ-neighborhood of ζ'. An infinite number of the z_n would then have a distance of more than ϵ from each other (the reader should make a little sketch). But that is impossible, because by hypothesis *nearly all* z_n are supposed to have a distance of less than ϵ from each other. Hence, ζ is the only limit point, and $z_n \to \zeta$, Q.E.D.

Every sequence of numbers which does *not* converge is called *divergent*. If a sequence converges to 0, $z_n \to 0$, it is called a *null sequence*.

There are the following simple, but important, theorems regarding operations on convergent sequences, which are proved exactly as in the real domain:

THEOREM 2. *Let* $\{z_n\}$ *and* $\{z_n'\}$ *be sequences such that* $z_n \to \zeta$ *and* $z_n' \to \zeta'$. *If* c, c' *are two arbitrary complex numbers, then the sequence* $\{w_n\}$ *with the terms* $w_n = cz_n + c'z_n'$ *is also convergent, and* $w_n \to c\zeta + c'\zeta'$.

THEOREM 3. *Under the same hypotheses as in the preceding theorem, the sequence* $\{w_n\}$ *with the terms* $w_n = z_n z_n'$ *is also convergent, and* $w_n \to \zeta\zeta'$.

THEOREM 4. *If $z_n \to \zeta$, all $z_n \neq 0$, and $\zeta \neq 0$, then the sequence $\{w_n\}$ with the terms $w_n = 1/z_n$ is also convergent, and $w_n \to 1/\zeta$.*

THEOREM 5. *If $z_n \to \zeta$, and if $\{z_n'\}$ is a subsequence[51] of the sequence $\{z_n\}$, then also $z_n' \to \zeta$.*

27. Sequences of real numbers

If all the terms of a sequence of numbers are real, it is called, briefly, a *real sequence*. Since such sequences are special cases of "complex" sequences, the considerations of the preceding paragraph are also valid for these real sequences $\{x_n\}$. Several new details arise here, however, in connection with §24:

A sequence $\{x_n\}$ which is bounded on the left has a well-determined *greatest lower bound* γ, which is characterized by the following two conditions: There is no $x_n < \gamma$; but, for an arbitrary $\epsilon > 0$, there is *at least one* $x_n < \gamma + \epsilon$. A corresponding statement holds for the least upper bound γ'.

It has also a well-determined lower limit μ, in symbols:

$$\liminf x_n = \mu \qquad \text{or} \qquad \underline{\lim} \, x_n = \mu,$$

which satisfies the following two conditions: For every $\epsilon > 0$,

$$\text{at most finitely many } x_n < \mu - \epsilon,$$

$$\text{but infinitely many } x_n < \mu + \epsilon.$$

And a corresponding statement holds for the lim sup x_n or $\overline{\lim} \, x_n = \mu'$. It is clear, according to §24, without further comment, when to set any one of the four numbers spoken of equal to $\pm \infty$.

Obviously the real sequence $\{x_n\}$ is convergent if, and only if, $\mu = \mu'$ and the common value is finite; this number is the limit of the sequence.

A real sequence $\{x_n\}$ is called *monotonically increasing*, if always $x_n \leqq x_{n+1}$; *monotonically decreasing*, if always $x_n \geqq x_{n+1}$. For such sequences we have the important

THEOREM 1. *A monotonically increasing sequence is convergent if, and only if, it is bounded on the right; a monotonically decreasing sequence, if, and only if, it is bounded on the left.*

---

[51]If $k_1, k_2, \ldots, k_n, \ldots$ is any increasing sequence of natural numbers, then the sequence $\{z_n'\}$ with the terms $z_n' = z_{k_n}$ is called a *subsequence* of $\{z_n\}$.

For, its lim sup (or its lim inf) is obviously its only limit point. It is therefore also the limit.

Now we can give the Dedekind cut a form which is often handier for applications: Let $\{a_n\}$ be a real, *monotonically increasing* sequence, and $\{a_n'\}$ a real, *monotonically decreasing* sequence. Moreover, let always $a_n < a_n'$. Finally, let $a_n' - a_n = l_n \to 0$. Since the interval $\langle a_n, a_n' \rangle$ then contains the next interval $\langle a_{n+1}, a_{n+1}' \rangle$, we say, for brevity, that we have a *nest of intervals*. (The lengths, l_n , of its intervals form a null sequence.) Concerning it, there is the following *principle of nested intervals*:

THEOREM 2. *There is always one, and only one, point which belongs to all the intervals of a nest of intervals.*

For, according to the preceding theorem, the sequences $\{a_n\}$ and $\{a_n'\}$ are convergent. Let λ and λ' be their respective limits. Then, moreover, for every n,

$$a_n \leqq \lambda \leqq \lambda' \leqq a_n' \quad \text{and} \quad \lambda' - \lambda \leqq l_n .$$

Hence, $\lambda' = \lambda$, and this point belongs to all the intervals. A number λ^* distinct from λ, however, cannot also belong to all the intervals; otherwise the length of *every* interval would have to be at least equal to the (positive) distance between λ and λ^*, whereas the lengths l_n were supposed to tend to 0.

The matters dealt with in this paragraph and in §24 belong exclusively to the theory of *real* numbers and *real* sets. We were therefore justified in assuming that the reader is already familiar with them in the main, and in expressing ourselves briefly, accordingly. The same holds for the theory of infinite series in the real domain, which is treated in the next paragraph.

28. Infinite series

A sequence is very often given *indirectly* by supposing a first sequence $\{c_n\}$ to be given directly, and obtaining from it a new sequence, $\{s_n\}$, by stipulating that

$$s_0 = c_0 , \qquad s_1 = c_0 + c_1 , \qquad s_2 = c_0 + c_1 + c_2 ,$$

and, in general,

$$(1) \qquad s_n = c_0 + c_1 + \cdots + c_n , \qquad (n = 0, 1, 2, \ldots).$$

The sequence $\{s_n\}$ is then denoted briefly by the symbol

(2) $$\sum_{n=0}^{\infty} c_n \quad \text{or simply} \quad \sum c_n \, ,$$

and is called an *infinite series*. The c_n are named its *terms*; the s_n, its partial sums. The symbol (2) thus denotes the sequence of partial sums (1). If the latter is convergent (or divergent), the series (2) is also called convergent (divergent). In the first case, the limit, s, of s_n is designated as the *value* or the *sum* of the series (2). The symbol (2) is then also used as a symbol for the number s itself:

(3) $$\sum_{n=0}^{\infty} c_n = s.$$

Cauchy's convergence principle of §26 now yields immediately

THEOREM 1. *The series $\sum c_n$ is convergent if, and only if, after having chosen $\epsilon > 0$, an $n_0 = n_0(\epsilon)$ can always be assigned so that*

(4) $$| c_{n+1} + c_{n+2} + \cdots + c_{n+p} | < \epsilon$$

for all $n > n_0$ and all $p > 0$.

This theorem leads at once to the following two:

THEOREM 2. *The terms of a convergent series $\sum c_n$ form a null sequence:* $c_n \to 0$.

Because, for $p = 1$, (4) asserts that $| c_{n+1} | < \epsilon$ for all n after a certain one.

THEOREM 3. *If the series $\sum | c_n |$ (which has non-negative real terms) converges, then the series $\sum c_n$ is also convergent.*

Because, we have always (see §13, 1).

$$| c_{n+1} + c_{n+2} + \cdots + c_{n+p} | \leqq | c_{n+1} | + | c_{n+2} | + \cdots + | c_{n+p} |.$$

All these things are formally the same as in the real domain; even the definition of *absolute convergence*:

DEFINITION. *A series $\sum c_n$ is called absolutely convergent, if the series $\sum | c_n |$ converges. If $\sum c_n$ is convergent, but $\sum | c_n |$ is not, then $\sum c_n$ is called, more precisely, conditionally (or "only conditionally") convergent.*

In virtue of this definition and Theorem 3, the question of the *absolute* convergence of series of complex terms is completely reduced to a convergence question for series of non-negative real terms. We can therefore formulate immediately the following important criteria for the absolute convergence of a series $\sum c_n$ of complex terms, since they are simply the well-known convergence tests for the real series $\sum |c_n|$.

I. *Convergence criteria for the absolute convergence of series of complex terms*:

1. $\sum c_n$ *is absolutely convergent if, and only if, the sequence of real, monotonically increasing numbers,*

$$\sigma_n = |c_0| + |c_1| + \cdots + |c_n|,$$

is bounded.

The following criteria have the character of only necessary conditions. First, from 1. follows immediately the so-called *comparison test*:

2. *If $\sum \gamma_n$ is a convergent series of positive real terms, and if always $|c_n| \leqq \gamma_n$, then the series $\sum c_n$ is absolutely convergent.*

The following two are particularly important for application:

3. $\sum c_n$ *is absolutely convergent, if there exists a positive number $\gamma < 1$, such that nearly all quotients*

$$(5) \qquad \left| \frac{c_{n+1}}{c_n} \right| \leqq \gamma.$$

Or we may say

3′. $\sum c_n$ *is absolutely convergent, if*

$$(6) \qquad \overline{\lim} \left| \frac{c_{n+1}}{c_n} \right| = \lambda < 1.$$

(For if (5) is satisfied, then also $\lambda \leqq \gamma < 1$. And if $\lambda < 1$, then (5) is satisfied with, e.g., $\gamma = \frac{1}{2}(\lambda + 1) < 1$.)

4. $\sum c_n$ *is absolutely convergent, if there exists a positive number $\gamma < 1$, such that nearly all radicals*

$$(7) \qquad \sqrt[n]{|c_n|} \leqq \gamma.$$

For the same reason as in 3., we may express this as follows:

4′. $\sum c_n$ *is absolutely convergent, if*

$$(8) \qquad \overline{\lim} \sqrt[n]{|c_n|} = \lambda < 1.$$

It is divergent, if $\lambda > 1$ (because then the terms of the series do not form a null sequence).

The criteria 3. are called *ratio tests*; the criteria 4., *radical tests*.

II. The *rules for operating with convergent series* are also formally the same as in the real domain, and follow likewise from the corresponding rules for operating with convergent sequences of numbers:

1. *If* $\sum c_n$ *and* $\sum c'_n$ *are two convergent series with the respective sums s and s', and if c and c' are two arbitrary complex numbers, then the series*

$$\sum (cc_n + c'c'_n)$$

is also convergent, and its sum is equal to $cs + c's'$.

(Proof according to §26, Theorem 2.) We say that convergent series *may be multiplied (term by term) by a constant*, and *may be added term by term*.

Let $\{k_n\}$ be a sequence of natural numbers, in which every natural number (possibly also 0) appears once, and only once. Then the series $\sum c'_n$, with $c'_n = c_{k_n}$, is called a *rearrangement* of the series $\sum c_n$.

2. *If the series* $\sum c_n$ *is absolutely convergent and has the sum s, then every one of its rearrangements,* $\sum c'_n$, *is absolutely convergent and has the same sum s.*[52]

Proof: Let $\epsilon > 0$ be given arbitrarily. Then, by hypothesis, m can be chosen so that

$$(9) \qquad |c_{m+1}| + |c_{m+2}| + \cdots + |c_{m+p}| < \epsilon$$

for every p. We now choose n_0 so large, that all the numbers $0, 1, \ldots, m$ are contained among the numbers $k_0, k_1, \ldots, k_{n_0}$. Denote by s'_n the partial sums of $\sum c'_n$. Then, for $n > n_0$, all the terms $c_0, c_1, \ldots, c_m$ are annulled in the difference $s'_n - s_n$, and only a finite number of terms remain, whose sum, because of (9), is certainly less than ϵ. Hence, for $n > n_0$, $|s'_n - s_n| < \epsilon$, so that $(s'_n - s_n) \to 0$. But then, since $s'_n = s_n + (s'_n - s_n)$, the sequence $\{s'_n\}$ has the same limit as the

[52] We add, without proof, that this theorem does *not* hold for series which converge only conditionally.

sequence $\{s_n\}$; i.e., $\sum c_n'$ is also convergent, and has the sum s. If the same proof is repeated with the two series $\sum |c_n|$ and $\sum |c_n'|$, it is seen that the latter, too, converges, and, consequently, $\sum c_n'$ is absolutely convergent.

3. Let $\sum c_n$ and $\sum c_n'$, now, be any two infinite series. We form the products

$$c_k c_l', \qquad (k = 0, 1, 2, \cdots; l = 0, 1, 2, \cdots),$$

of every single term of the first series and every single term of the second. These products can be arranged in the most varied manners to form a simple sequence $\{p_n\}$. To this end, first arrange the products as in a determinant (k = row number, l = column number):

$$
\begin{array}{llll}
c_0 c_0', & c_0 c_1', & c_0 c_2', & \cdots \\
c_1 c_0', & c_1 c_1', & c_1 c_2', & \cdots \\
c_2 c_0', & c_2 c_1', & c_2 c_2', & \cdots \\
\multicolumn{4}{c}{\cdots\cdots\cdots\cdots\cdots\cdots\cdots}
\end{array}
$$

(10)

The *arrangement by diagonals* is then obtained by writing down the products for which $k + l$ has the successive values 0, 1, 2, . . . , running through each diagonal from top to bottom, say. We get the *arrangement by squares*, if we take, in succession, the squares which correspond to these diagonals; i.e., those products for which k and $l = 0, \leqq 1, \leqq 2, \ldots$.

Every series $\sum p_n$ obtained in such a manner is called a *product series* of the two series $\sum c_n$ and $\sum c_n'$, and there is the following theorem concerning it:

If the series $\sum c_n$ and $\sum c_n'$ are both absolutely convergent, and s and s' are their respective sums, then every one of their product series is absolutely convergent and has the sum ss'.

Proof: Obviously

$$|p_0| + |p_1| + \cdots + |p_n|$$
$$\leqq (|c_0| + \cdots + |c_m|)(|c_0'| + \cdots + |c_m'|),$$

provided m is taken sufficiently large. The sequence of partial sums of $\sum |p_n|$ is thus bounded, and, consequently, $\sum p_n$ is absolutely convergent. By 2., then, all product series have the same sum S, since they are merely rearrangements of each other. If $\{p_n\}$ denotes, in particular, the arrangement by squares, we have

$$(c_0 + c_1 + \cdots + c_n)(c_0' + c_1' + \cdots + c_n')$$
$$= p_0 + p_1 + \cdots + p_{(n+1)^2-1} .$$

From this it follows (with the aid of Theorems 3 and 5 of §26), on letting $n \to \infty$, that $S = ss'$.

4. Let us group sets of successive terms of a convergent infinite series $\sum c_n$, with sum s, by means of parentheses, and regard each set as a new term; i.e., form the series

$$(c_0 + c_1 + \cdots + c_{k_0}) + (c_{k_0+1} + \cdots + c_{k_1})$$
$$+ (c_{k_1+1} + \cdots + c_{k_2}) + \cdots ,$$

and call its terms C_0, C_1, ... We say that $\sum C_n$ is obtained from $\sum c_n$ by *grouping terms*. There is then the following theorem:

Let $\sum c_n$ be convergent and equal s. Then every series, $\sum C_n$, obtained from $\sum c_n$ by grouping terms, is also convergent and has the same sum s.

For, the sequence of partial sums of $\sum C_n$ is obviously a subsequence of the sequence of partial sums of $\sum c_n$.

Evidently, if $\sum c_n$ is absolutely convergent, so is $\sum C_n$.

5. One has especially frequent occasion to form such a grouping of terms of that product series of two series, $\sum c_n$ and $\sum c_n'$, which is obtained from the arrangement *by diagonals*. If the products in the same diagonal are combined by means of parentheses, thus forming the series

$$(11) \qquad \sum_{n=0}^{\infty} (c_0 c_n' + c_1 c_{n-1}' + \cdots + c_n c_0'),$$

this is called the *Cauchy product* of the given series. From the last two theorems, there follows:

The Cauchy product of two absolutely convergent series is again absolutely convergent, and we have

$$(12) \qquad \sum_{n=0}^{\infty} (c_0 c_n' + c_1 c_{n-1}' + \cdots + c_n c_0') = \left(\sum_{n=0}^{\infty} c_n \right)\left(\sum_{n=0}^{\infty} c_n' \right).$$

Examples of these matters are contained in the next chapter and in all of section 5.

We finally add, without proof, the following somewhat more far-reaching theorem. Its proof, as that of every one of the preceding theorems, is exactly the same as in the real domain.

6. Let there be given, as in (10), an infinite matrix of the form

$$(13) \qquad (c_{kl}) = \begin{pmatrix} c_{00} & c_{01} & c_{02} & \cdots \\ c_{10} & c_{11} & c_{12} & \cdots \\ c_{20} & c_{21} & c_{22} & \cdots \\ \cdots\cdots\cdots\cdots\cdots \end{pmatrix}.$$

If its elements are arranged in any manner to form a simple sequence $\{c_n\}$, and if $\sum. c_n$ is absolutely convergent, then all "row series"

$$(14) \qquad \sum_{l=0}^{\infty} c_{kl} = Z_k , \qquad (k = 0, 1, 2, \cdots),$$

and all "column series"

$$(15) \qquad \sum_{k=0}^{\infty} c_{kl} = S_l , \qquad (l = 0, 1, 2, \cdots),$$

are also absolutely convergent. The same is true of the series $\sum Z_k$ and $\sum S_l$, and we have

$$(16) \qquad \sum_{k=0}^{\infty} Z_k = \sum_{l=0}^{\infty} S_l = \sum c_n .$$

(*Cauchy's double-series theorem.*)

CHAPTER VIII

POWER SERIES

29. *The circle of convergence*

Those series $\sum c_n$ are of particular importance for the theory of functions, for which c_n has the form $a_n(z - z_0)^n$; that is, the series

$$
(1) \quad \sum_{n=0}^{\infty} a_n(z - z_0)^n
$$
$$
\equiv a_0 + a_1(z - z_0) + \cdots + a_n(z - z_0)^n + \cdots .
$$

Such a series is called a *power series* with the "center" z_0 and the "coefficients" a_n. We think of z_0 and the a_n as given, and the question is: For what values of z is the given series convergent, for what values not?

Examples. 1. $z_0 = 0$, all $a_n = 1$. This gives the so-called *geometric series*,

$$
(2) \quad \sum_{n=0}^{\infty} z^n = 1 + z + z^2 + \cdots + z^n + \cdots .
$$

The comparison test, radical test, or ratio test, shows that this series converges absolutely for $|z| < 1$. For $|z| \geqq 1$ it is divergent, because then the sequence of terms does not tend to 0. Thus, the geometric series is absolutely convergent precisely in the interior of the unit circle, divergent everywhere else. Since, moreover, $z^n \to 0$ for $|z| < 1$, we have for the partial sums:

$$
1 + z + \cdots + z^n = \frac{1 - z^{n+1}}{1 - z} = \frac{1}{1 - z} - \frac{z}{1 - z} z^n \to \frac{1}{1 - z}.
$$

Hence, the linear function $1/(1 - z)$ is represented by the geometric series in the interior of the unit circle:

$$
\sum_{n=0}^{\infty} z^n = \frac{1}{1 - z}, \qquad (|z| < 1).
$$

2. The power series

$$(3) \quad (z-1) + \frac{(z-1)^2}{2} + \cdots + \frac{(z-1)^n}{n} + \cdots \equiv \sum_{n=1}^{\infty} \frac{(z-1)^n}{n}$$

is, as is just as easily ascertained, absolutely convergent for all z of the open circular disk $|z-1| < 1$. It is divergent for those z with $|z-1| > 1$. We leave open the question of convergence at the boundary points $|z-1| = 1$.

3. The power series

$$(4) \quad 1 + z + \frac{z^2}{2!} + \cdots + \frac{z^n}{n!} + \cdots \equiv \sum_{n=0}^{\infty} \frac{z^n}{n!}$$

is, as is shown immediately by the ratio test, *absolutely convergent for all z.* It is therefore called *everywhere convergent.* Further details concerning this series appear in chapter 12.

4. The power series $\sum n^n z^n$ can converge for no $z \neq 0$, because for a $z \neq 0$ the terms of the series do not form a null sequence. Such a power series is called *nowhere convergent.*

The typical behavior of arbitrary power series is revealed already by these examples, for we have the

FUNDAMENTAL THEOREM. *Let $\sum a_n(z-z_0)^n$ be a power series which is neither everywhere convergent nor nowhere convergent. Then there exists a definite positive number r such that the series converges absolutely at all points of the open circular disk $|z-z_0| < r$, but diverges at all points z with $|z-z_0| > r$. At the boundary points, $|z-z_0| = r$, it may converge or diverge.*[53]

The circle $|z-z_0| < r$ is therefore called, briefly, the *circle of convergence*, its radius, the *radius of convergence*, of the series. If it is everywhere convergent, we set $r = +\infty$; if it is nowhere convergent, $r = 0$.

We furnish the *proof* by showing at the same time:

COROLLARY. *The radius of convergence of the power series $\sum a_n(z-z_0)^n$ has the value*

$$(5) \qquad\qquad r = \frac{1}{\overline{\lim} \sqrt[n]{|a_n|}}.$$

[53]These points will generally be left out of consideration in what follows.

Or, more precisely: *If we set* $\overline{\lim} \sqrt[n]{|a_n|} = \mu$, *then*

$$r = \begin{cases} 1/\mu \\ +\infty \\ 0 \end{cases} \quad according\ as \quad \begin{cases} 0 < \mu < +\infty \\ \mu = 0 \\ \mu = +\infty \end{cases}$$

For suppose $0 < \mu < +\infty$. Then, for fixed z,

$$\overline{\lim} \sqrt[n]{|a_n(z - z_0)^n|} = |z - z_0|\, \mu.$$

Hence, by the radical test, the power series is absolutely convergent for $|z - z_0| < r < 1/\mu$, divergent for $|z - z_0| > r$.

If $\mu = 0$, let $\epsilon > 0$ be chosen arbitrarily. Then nearly all values $\sqrt[n]{|a_n|} < \epsilon$. If $z_1 \neq z_0$ is an arbitrary point of the plane, nearly all values $\sqrt[n]{|a_n|} < 1/2|z_1 - z_0|$. Consequently, for nearly all terms of our power series, we have

$$|a_n(z_1 - z_0)^n| < 1/2^n.$$

It is therefore absolutely convergent at $z = z_1$, by the comparison test. Since this is, of course, also the case at $z = z_0$, it is *everywhere convergent*.

If, however, $\mu = +\infty$, then, for every point $z_1 \neq z_0$, we have

$$\overline{\lim} \sqrt[n]{|a_n(z_1 - z_0)^n|} = +\infty,$$

so that the power series is divergent at z_1. It is thus *nowhere convergent*.

Since the three cases considered are mutually exclusive, and further cases do not exist (because $\sqrt[n]{|a_n|} \geqq 0$), it follows, in addition, that the conditions stated for their occurrence are not only sufficient, but also necessary.

The power series

$$a_1 + 2a_2(z - z_0) + \cdots + na_n(z - z_0)^{n-1} + \cdots$$

(6)
$$\equiv \sum_{n=1}^{\infty} na_n(z - z_0)^{n-1} \equiv \sum_{n=0}^{\infty} (n + 1)a_{n+1}(z - z_0)^n$$

is called the (formal) derivative of $\sum a_n(z - z_0)^n$. Due to the fact that $\sqrt[n]{n} \to 1$, the former has the same radius of convergence as the latter. The same is also true of the formal integral:

$$(7) \quad a_0(z - z_0) + \frac{a_1}{2}(z - z_0)^2 + \cdots + \frac{a_n}{n+1}(z - z_0)^{n+1} + \cdots .$$

30. *Operations on power series*

In our further considerations, we shall assume that $z_0 = 0$, which is obviously no restriction.[54] According to Rule II, 1 in §28, it then follows immediately, that a power series *may be multiplied term by term by a constant*, and that two power series *may be added term by term*,

$$(1) \qquad \sum a_n z^n + \sum b_n z^n = \sum (a_n + b_n)z^n,$$

provided that for each of the series, z lies in the *interior* of the circle of convergence. Under the same conditions, we may also form their *Cauchy product* (§28, II, 5):

$$(2) \quad (\sum a_n z^n)(\sum b_n z^n) = \sum (a_0 b_n + a_1 b_{n-1} + \cdots + a_n b_0)z^n,$$

and we see that this kind of product formation is of particular importance for power series. By repeated application, it follows that the powers of a power series, $(\sum a_n z^n)^2$, $(\sum a_n z^n)^3$, etc., can be represented as power series, so long as z lies in the interior of the circle of convergence. We set[55]

$$(3) \qquad (\sum a_n z^n)^k = \sum a_n^{(k)} z^n, \qquad (k = 1, 2, 3, \cdots).$$

To master the *division* of power series, it suffices to represent the reciprocal value of a power series,

$$\frac{1}{a_0 + a_1 z + a_2 z^2 + \cdots} = \frac{1}{a_0} \frac{1}{1 + \frac{a_1}{a_0} z + \frac{a_2}{a_0} z^2 + \cdots},$$

again as a power series, assuming that $a_0 \neq 0$. If we set $a_n/a_0 = -b_n$, the problem is to represent

$$(4) \qquad \frac{1}{1 - (b_1 z + b_2 z^2 + \cdots)}$$

[54]For we can set $(z - z_0) = z'$ and then drop the accent.

[55]We shall not investigate the formation of the coefficients $a_n^{(k)}$ for larger values of k; it is not important for what follows.

as a power series: If we place

(5) $b_1 z + b_2 z^2 + \cdots = w, \ |b_1 z| + |b_2 z^2| + \cdots = W$

and if $|W| < 1,$[56] and hence also $|w| < 1$, then, according to §29, Example 1,

(6) $$\frac{1}{1-w} = 1 + w + w^2 + \cdots .$$

By virtue of (3), we obtain from (5) the expansions

$$w = b_1 z + b_2 z^2 + \cdots$$

(7) $$w^2 = b_1^{(2)} z + b_2^{(2)} z^2 + \cdots$$

$$w^3 = b_1^{(3)} z + b_2^{(3)} z^2 + \cdots$$
$$\cdots\cdots\cdots\cdots\cdots\cdots\cdots\cdots .$$

Since all the series used thus far are also convergent if the a_n and b_n and z are replaced by their respective absolute values, the *column series* in this array are (absolutely) convergent; and if we set

$$b_l + b_l^{(2)} + b_l^{(3)} + \cdots = c_l ,$$

then (according to §28, II, 6)

(8) $$\sum_{l=1}^{\infty} c_l z^l = \text{(the sum of the row series)} = \sum_{k=1}^{\infty} w^k.$$

Hence, with the coefficients c_l thus calculated,

(9) $$\frac{1}{1 - (b_1 z + b_2 z^2 + \cdots)} = 1 + c_1 z + c_2 z^2 + \cdots ,$$

provided that z lies in the interior of the circle of convergence of the series (5), and that the sum of the second one of these series is less than $1.$[57] (For a more convenient calculation of the coefficients, cf. §41, 9.)

By means of a trivial generalization of the last consideration,

[56]In §35 we shall see that this is the case for all points z sufficiently close to 0.

[57]This too is automatically the case for all points lying sufficiently close to 0.

one can prove, finally, the following more far-reaching

THEOREM. *Let*

$$(5') \qquad w = b_1 z + b_2 z^2 + \cdots$$

be a power series with positive radius r, and let

$$(6') \qquad \mathfrak{w} = c_0 + \beta_1 w + \beta_2 w^2 + \cdots$$

be another power series, with positive radius R. Then, for all z less than r in absolute value, and such that

$$(5'') \qquad |b_1 z| + |b_2 z^2| + \cdots < R,^{58}$$

we may "substitute" the first series in the second. That is to say: If we form equations (7) as before, and multiply these, in succession, by β_1 , β_2 , $\cdots$, then the column series of the resulting array (7') are convergent. And if we set their respective sums equal to $c_1 z, c_2 z^2, \ldots$, then the power series

$$(8') \qquad c_0 + c_1 z + c_2 z^2 + \cdots$$

is also convergent for the aforementioned z.

With the use of concepts which are explained more precisely in the next chapter, we can add to this: *If (5') represents the function $f(z)$, (6') the function $g(w)$, then the power series (8') represents the composite function $g(f(z))$.*

[58]This too is automatically the case for all points lying sufficiently close to 0.

CHAPTER IX

FUNCTIONS OF A COMPLEX VARIABLE

31. *The concept of a function of a complex variable*

The concept of a function is defined in the complex domain formally the same as in the real domain:

If $\mathfrak{M}$ is an arbitrary point set, and if z is allowed to denote any point of $\mathfrak{M}$, then z is called a (complex) *variable*, and $\mathfrak{M}$ is called the *domain of variation* of z. Now, if there exists a rule by virtue of which a certain new number w is made to correspond to every point z of $\mathfrak{M}$, w is called a (*single-valued*) *function of the* (*complex*) *variable z*; in symbols:

$$(1) \qquad w = f(z),$$

where f (or any other suitable letter, such as F, g, h, etc.) stands for the prescribed rule. $\mathfrak{M}$ is called the *domain of definition* of the function, and z, its *argument*. The totality of values w which correspond to the points z of $\mathfrak{M}$, is called the *domain of values* of the function (over $\mathfrak{M}$).

In the following, we shall consider only the case that $\mathfrak{M}$ is a circular region or the entire plane (sometimes with the exclusion of certain points) and that the functional value $w = f(z)$ is given by means of some simple closed expression or as the sum of a power series.

In section II we became acquainted with the *linear functions*. The *rational functions* constitute an obvious generalization: these are the functions whose defining rule combines the variable z with any constants by means of the *rational* operations, and they can therefore be expressed in the form

(2)
$$\frac{b_0 + b_1 z + b_2 z^2 + \cdots b_q z^q}{a_0 + a_1 z + \cdots + a_p z^p}.$$

If only the *integral* operations are employed, the expression is called an *entire* rational function, and it can be expressed in the form

(3)
$$a_0 + a_1 z + a_2 z^2 + \cdots + a_p z^p.$$

These and the other so-called "elementary functions" are discussed in somewhat greater detail in section V. Every power series defines, in its circle of convergence, a function of the complex variable z.

The geometric interpretation of such a function of a complex argument is considered in the next chapter.

32. *Limits of functions*

Let ζ be an *interior* point of the domain of definition of a function $w = f(z)$. Then we say—formally precisely as in the real domain—that

(1) $\qquad f(z) \to \omega \qquad$ (or $w \to \omega$) $\qquad$ as $z \to \zeta$,

or

(2)
$$\lim_{z \to \zeta} f(z) = \omega,$$

if one of the following two conditions is satisfied:[59]

1) Having chosen $\epsilon > 0$, it is possible to assign a $\delta = \delta(\epsilon) > 0$ such that

(3) $\qquad |f(z) - \omega| < \epsilon, \qquad (0 < |z - \zeta| < \delta),$

for all values of z which belong to the domain of definition of $f(z)$ and satisfy the condition $0 < |z - \zeta| < \delta$.

2) For every sequence of numbers $\{z_n\}$ which approaches the limit ζ and whose terms, all different from ζ, are taken from the

---

[59]Since no mention is made, in the following conditions, of a functional value at the point $z = \zeta$ itself, $f(z)$ need not be defined at ζ.

domain of definition of $f(z)$, the sequence of the corresponding functional values $w_n = f(z_n)$ approaches[60] the limit ω.

These two conditions are fully equivalent. It is obvious that the second is satisfied if the first is, because nearly all z_n lie in the δ-neighborhood of ζ if $z_n \to \zeta$. Conversely, if the first condition is *not* satisfied, this means that an $\epsilon_0 > 0$ exists having the property that points, z, for which $| f(z) - \omega | \geqq \epsilon_0$, lie in *every* neighborhood of ζ. But then one can also assign a sequence $\{z_n\}$ of such z-values, which approaches ζ, but for which $f(z_n)$ does *not* approach ω.

Since the definition of (1) and (2) is formally precisely the same as in the real domain, the same rules hold for *operations on limits of functions* in the complex and real domains:

I. If ζ is an interior point of each of the domains of definition of the two functions $f_1(z)$ and $f_2(z)$, it is also an interior point of the domain of definition of the function

$$c_1 f_1(z) + c_2 f_2(z),$$

where c_1, c_2 denote any two complex numbers. Now, if, as $z \to \zeta$,

$$f_1(z) \to \omega_1 \qquad \text{and} \qquad f_2(z) \to \omega_2,$$

then

(4) $$f(z) = c_1 f_1(z) + c_2 f_2(z) \to c_1 \omega_1 + c_2 \omega_2 ;$$

in particular, $f_1(z) \pm f_2(z) \to \omega_1 \pm \omega_2$, and $c_1 f_1(z) \to c_1 \omega_1$.

II. Under the same hypotheses as in I.,

$$f_1(z) \cdot f_2(z) \to \omega_1 \omega_2.$$

III. Under the same hypotheses, *and if $\omega_2 \neq 0$,*

$$\frac{f_1(z)}{f_2(z)} \to \frac{\omega_1}{\omega_2}$$

as $z \to \zeta$.

On the basis of definition 2), moreover, these rules follow directly from the corresponding rules for operating with convergent sequences of numbers (§26).

[60]Whereas in the real domain, the variable can approach a particular point only from the right or the left (or from both sides), in the complex domain, z or z_n can approach ζ *from all directions.*

Examples. 1) If a function $w = f(z)$ is defined by setting, for *every z*, the corresponding w equal to one and the same constant, c, we say that $f(z)$ is *identically equal to c*, or is *identically constant*. For this function we have, of course, also $\lim f(z) = c$ as $z \to \zeta$, and this holds for every point ζ.

2) Let $f(z) = z$ for every z. Then for this function obviously

$$\lim_{z \to \zeta} f(z) = \zeta$$

for every point ζ.

3) By repeated application of II., it follows, now, that, for every point ζ, the non-negative integral power z^k has the limit ζ^k as $z \to \zeta$:

$$z^k \to \zeta^k \qquad \text{as} \qquad z \to \zeta.$$

4) By applying I. and II., it follows, from these examples, that if $f(z) = a_0 + a_1 z + \cdots + a_p z^p$ is an arbitrary entire rational function, then the relation

$$\lim_{z \to \zeta} f(z) = f(\zeta)$$

is valid for every ζ.

5) From this it follows, finally, by applying III., that if $g(z) = b_0 + b_1 z + \cdots + b_q z^q$ is a second entire rational function, the relation

$$\lim_{z \to \zeta} \frac{g(z)}{f(z)} = \frac{g(\zeta)}{f(\zeta)}$$

holds for every ζ *for which* $f(\zeta) \neq 0$. Thus, the limit of a rational function, at every point at which its denominator does not vanish, is equal to the value of the function at that point.

33. Continuity

The situation encountered in the last example is of particular importance. We set it down in the form of a special definition:

DEFINITION. 1) *A function $f(z)$ is said to be continuous at an interior point ζ of its domain of definition, if*

$$\lim_{z \to \zeta} f(z) = f(\zeta).$$

If we employ the two definitions of limit given in §32, we can also say that $f(z)$ is said to be continuous at the point ζ if one of the following two (equivalent) conditions is satisfied:

2) To every $\epsilon > 0$, it is possible to make correspond a $\delta = \delta(\epsilon) > 0$, such that

$$| f(z) - f(\zeta) | < \epsilon$$

for all z with $| z - \zeta | < \delta$.[61]

3) For *every* sequence of numbers, $\{z_n\}$, chosen from the domain of definition of $f(z)$ in such a manner that $z_n \to \zeta$, we have, for the corresponding functional values: $f(z_n) \to f(\zeta)$.

In view of the examples of the preceding paragraph, we can say immediately: A rational function is continuous at every point at which its denominator does not vanish. The entire rational functions are everywhere continuous.

Likewise, from Rules I–III (§32), there follows immediately:

If the two functions $f_1(z)$ and $f_2(z)$ are continuous at the point ζ, then the functions

$$c_1 f_1(z) + c_2 f_2(z), \; f_1(z) \cdot f_2(z), \text{ and } f_1(z)/f_2(z)$$

are also continuous at this point ζ,—the last, however, only if $f_2(\zeta) \neq 0$.

34. Differentiability

We finally carry over to functions of a complex argument the concept which is in many respects the most important, that of *differentiability*:

DEFINITION. 1) *A function* $w = f(z)$ *is said to be differentiable at an interior point* ζ *of its domain of definition, if the limit*

$$(1) \qquad \lim_{z \to \zeta} \frac{f(z) - f(\zeta)}{z - \zeta} \qquad or \qquad \lim_{z \to \zeta} \frac{w - \omega}{z - \zeta}, \qquad (\omega = f(\zeta)),$$

exists. Its value is called the derivative or the differential quotient of the function $f(z)$ *at the point* ζ, *and it is denoted by*

[61]In other words: For a z lying (sufficiently) close to ζ, the functional values $f(z)$ and $f(\zeta)$ differ by an arbitrarily small amount.

(2) $\qquad f'(\zeta), \qquad \dfrac{df(z)}{dz}, \qquad \dfrac{dw}{dz}, \qquad w'.$ [62]

If we make use of the two conditions for the existence of a limit given in §32, we can also say that a function $w = f(z)$ is said to be differentiable at the point ζ if one of the following two conditions is fulfilled:

2) A number ω' exists with the property that to every $\epsilon > 0$ it is possible to make correspond a $\delta = \delta(\epsilon) > 0$ such that

$$\left| \frac{f(z) - f(\zeta)}{z - \zeta} - \omega' \right| < \epsilon$$

for all z with $0 < |z - \zeta| < \delta$.

3) There exists a number ω' such that, for *every* sequence of numbers, $\{z_n\}$, chosen from the domain of definition of $f(z)$, tending to ζ as a limit, and having all its terms different from ζ, the sequence of difference quotients

$$\frac{w_n - \omega}{z_n - \zeta} \to \omega'.$$

Since this definition of differentiability is formally exactly the same as in the real domain, and since operations with sequences and limits are performed exactly as there, the proofs of the following fundamental rules of the differential calculus are also precisely the same as in the real domain:

If the two functions $f_1(z)$ and $f_2(z)$ are differentiable at the point ζ, then

I. the function $f(z) = c_1 f_1(z) + c_2 f_2(z)$ is differentiable at the point ζ, and we have

$$f'(\zeta) = c_1 f_1'(\zeta) + c_2 f_2'(\zeta);$$

II. the function $f(z) = f_1(z) \cdot f_2(z)$ is differentiable at the point ζ, and we have

$$f'(\zeta) = f_1'(\zeta) f_2(\zeta) + f_1(\zeta) f_2'(\zeta);$$

III. the function $f(z) = f_2(z)/f_1(z)$ is differentiable at the point ζ, provided that $f_1(\zeta) \neq 0$, and we have

[62]In the last three notations, the point $z = \zeta$ must then be added besides, or be known from the context.

$$f'(\zeta) = \frac{f_1(\zeta)f_2'(\zeta) - f_2(\zeta)f_1'(\zeta)}{(f_1(\zeta))^2}.$$

Examples. 1) If $f(z)$ is identically equal to c, then obviously $f'(\zeta) = 0$ for every ζ.

2) If $f(z) = z$, it follows immediately from the definition, that $f'(\zeta) = 1$ at every point ζ.

3) By repeated application of Rule II, it follows now, that

$$\frac{d(z^k)}{dz} = k\zeta^{k-1}$$

for every natural number k and at every point ζ.

4) Further, by applying Rules I–III, it follows that the derivative of a rational function also exists at every point at which the denominator of the function does not vanish, and that this derivative is calculated according to the same rules as in the real domain. Thus, e.g., the linear function $\dfrac{az + b}{cz + d}$ has the derivative $\dfrac{ad - bc}{(c\zeta + d)^2}$ at the point $\zeta \neq -\dfrac{d}{c}$.

If the domain of definition, $\mathfrak{M}$, of a function, $f(z)$, is open (see §23), and if $f(z)$ is differentiable at *every* point z of this domain, then $f'(z)$ is again a function defined in $\mathfrak{M}$, and is called, briefly, the *derivative* of $f(z)$ in $\mathfrak{M}$. If $f'(z)$ is differentiable in $\mathfrak{M}$, we obtain the *second derivative*, $f''(z)$, of $f(z)$ in $\mathfrak{M}$, and similarly we arrive at the *derivatives of higher order*. A rational function has derivatives of every order. Their domain of definition is the entire plane, from which those points where the denominator of the function vanishes, have been deleted.

If $f_1(z)$ is a function with the domain of definition $\mathfrak{M}_1$, and if all values of this function lie in the domain of definition, $\mathfrak{M}_2$, of a second function, $f_2(z)$, then one can form the *composite function*,

$$f(z) = f_2(f_1(z)),$$

whose domain of definition is again $\mathfrak{M}_1$. For its differentiation we have, as in the real domain, the so-called *chain rule*

IV. $$f'(z) = f_2'(f_1(\zeta)) \cdot f_1'(\zeta),$$

provided the "inner" function, $f_1(z)$, is differentiable at the point ζ, and the "outer" function, $f_2(z)$, is differentiable at the point $\zeta_1 = f_1(\zeta)$.

35. Properties of functions represented by power series

Let there now be given an arbitrary power series,[63]

$$(1) \qquad a_0 + a_1z + a_2z^2 + \cdots \equiv \sum_{n=0}^{\infty} a_nz^n,$$

whose radius of convergence, r, is positive (or ∞).[64] At every interior point of its circle of convergence, it has a definite sum. This is, consequently, a function $f(z)$ defined in $|z| < r$ by the power series. We also say that it is *represented* by the power series, or that it is *developed* or *expanded* in the power series, and we write

$$(2) \qquad f(z) = \sum_{n=0}^{\infty} a_nz^n.$$

The properties of such functions represented by power series— these functions are the only important ones, as the further development of the theory of functions shows—are established by the following theorems:

THEOREM 1. *The function represented by* (2) *is continuous at* $z = 0$.

Proof: If ρ, with $0 < \rho < r$, is chosen arbitrarily, the series $\sum |a_n| \rho^{n-1}$ is convergent; call its sum K. If $\{z_n\}$, now, is any sequence of numbers which lie in the circle of convergence, are all different from 0, but tend to 0 as a limit, then nearly all $|z_\nu| < \rho$, and for these z_ν we have

$$|f(z_\nu) - a_0| \leq \sum_{n=1}^{\infty} |a_n| |z_\nu^n| \leq K |z_\nu|.$$

Hence, $f(z_\nu) \to a_0 = f(0)$, which proves the continuity of $f(z)$ at 0.

---

[63]As already remarked in §30, it is no restriction to assume that the center $z_0 = 0$. The following Theorems 1 to 7 are thus also valid, after suitable changes in wording, for functions which are represented by power series with an arbitrary center z_0.

[64]The nowhere convergent power series are excluded now, as before.

From this theorem we easily obtain the important *identity theorem for power series*:

THEOREM 2. *If the power series* $\sum a_n z^n$ *and* $\sum b_n z^n$ *are both convergent for* $|z| < \rho$, *and if they have the same sum for all these z, then the series are identical; i.e.,* $a_n = b_n$ *for* $n = 0, 1, 2, \ldots$.

Proof: First, from

$$a_0 + a_1 z + a_2 z^2 + \cdots = b_0 + b_1 z + b_2 z^2 + \cdots$$

it follows, for $z = 0$, that $a_0 = b_0$. Accordingly, at least for all z with $0 < |z| < \rho$, we have also

$$a_1 + a_2 z + \cdots = b_1 + b_2 z + \cdots.$$

If we let $z \to 0$ in this equation,[65] it follows, further, that $a_1 = b_1$. Analogous deductions now yield in succession the equations $a_n = b_n$ for $n = 2, 3, \ldots$.

Since it already suffices for these deductions if z runs through a sequence of points $\{z_\nu\}$ tending to 0, the proof shows, in addition, that, for the identity of the two power series, it is sufficient that their sums coincide at each of an infinite number of distinct points which cluster about 0 as a limit point.

This identity theorem asserts that one and the same function cannot be developed in two distinct ways in a power series: If it is at all developable in a power series (with center z_0), this is possible in only one manner.

THEOREM 3. *The function represented by* (2) *can also be developed in a power series about any other point,* z_1, *in the interior of the circle of convergence, as center. Thus, if* $|z_1| < r$, *there exists always one, and only one, power series,*

$$(3) \qquad \sum_{k=0}^{\infty} b_k (z - z_1)^k,$$

with positive radius of convergence r_1, *which has likewise the sum* $f(z)$ *at those points z common to both circles of convergence; in fact, we have*

$$(4) \qquad b_k = \sum_{n=0}^{\infty} \binom{n+k}{k} a_{n+k} z_1^n ,$$

[65]Here it is essential that, in the limiting process $z \to 0$, z does not have the point 0 to represent (see §32, footnote 59).

and the radius, r_1, is at least equal to $r - |z_1|$.

Proof: $z = z_1 + (z - z_1)$, and hence

$$
\begin{aligned}
f(z) &= \sum_{n=0}^{\infty} a_n [z_1 + (z - z_1)]^n \\
(5) \qquad &= \sum_{n=0}^{\infty} a_n \Bigg[z_1^n + \binom{n}{1} z_1^{n-1}(z - z_1) \\
&\qquad + \binom{n}{2} z_1^{n-2}(z - z_1)^2 + \cdots + \binom{n}{n}(z - z_1)^n \Bigg].
\end{aligned}
$$

If we imagine the terms of this series to be written down in such a manner that the terms containing the same power of $(z - z_1)$ form a column, we obtain an array as in §28, II, 6, whose row sums are then precisely the terms of series (5). The kth column series, on the other hand, has exactly the sum $b_k(z - z_1)^k$, provided b_k has the meaning assigned in (4). The cited theorem of §28 would then immediately establish the asserted equality

$$
(6) \qquad f(z) = \sum_{k=0}^{\infty} b_k(z - z_1)^k,
$$

if the hypotheses of this theorem were satisfied. This is, indeed, the case. For if we replace all elements in the acquired array by their respective absolute values, the nth row-sum equals $|a_n| \, [|z_1| + |z - z_1|]^n$. But the sum over these row sums, that is, the series

$$
\sum_{n=0}^{\infty} |a_n| \, [|z_1| + |z - z_1|]^n,
$$

is still convergent if only $|z_1| + |z - z_1| < r$ or $|z - z_1| < r - |z_1|$. Therewith all is proved, including the corollary that the development (6) has a radius $r_1 \geqq r - |z_1|$.

THEOREM 4. *The function represented by (2) is continuous at every interior point z_1 of its circle of convergence.*

Proof: In a neighborhood of z_1, $f(z)$ is also represented by the series (6), thus again by a power series. Since the latter represents a function which is continuous at the center, z_1, of the series, $f(z)$ is continuous at z_1 by Theorem 1.

THEOREM 5. *The function represented by (2) is differentiable* *at every point z_1 of its circle of convergence, and the derivative there can be obtained by term-by-term differentiation; i.e., we have*

$$(7) \qquad f'(z_1) = \sum_{n=1}^{\infty} na_n z_1^{n-1} = \sum_{n=0}^{\infty} (n + 1)a_{n+1} z_1^n .$$

Proof: According to (6),

$$\frac{f(z) - f(z_1)}{z - z_1} = b_1 + b_2(z - z_1) + \cdots .$$

The power series on the right represents a function which is continuous at z_1. From this equality follows, then, immediately the assertion:

$$f'(z_1) = b_1 = \sum_{n=0}^{\infty} (n + 1)a_{n+1} z_1^n .$$

THEOREM 6. *The function represented by (2) is differentiable arbitrarily often at every point z_1 of its circle of convergence, and we have*

$$f^{(k)}(z_1) = k!b_k = \sum_{n=0}^{\infty} (n + 1)(n + 2) \cdots (n + k)a_{n+k} z_1^n ,$$

or, written more clearly,

$$(8) \qquad \frac{1}{k!} f^{(k)}(z_1) = b_k = \sum_{n=0}^{\infty} \binom{n + k}{k} a_{n+k} z_1^n .$$

Proof: For every $|z| < r$, by Theorem 5, we have

$$f'(z) = \sum_{n=0}^{\infty} (n + 1)a_{n+1} z^n.$$

The derivative $f'(z)$ is thus again represented by a power series with the same radius. Hence, by the same Theorem 5,

$$f''(z) = \sum_{n=0}^{\infty} (n + 1)(n + 2)a_{n+2} z^n; \text{ etc.}$$

If, finally, we substitute in the series (6) the values for b_k obtained in (8), we get the so-called *Taylor series*, i.e.,

THEOREM 7. *The function represented by (2) can be represented,* *for a neighborhood of every (interior) point z_1 of its circle of convergence, by the power series*

$$f(z) = \sum_{k=0}^{\infty} \frac{f^{(k)}(z_1)}{k!} (z - z_1)^k.$$

The most significant examples of these theorems are given in section V.

CHAPTER X

ANALYTIC FUNCTIONS AND CONFORMAL MAPPING

36. Analytic functions

While the preceding matters represent an exact transference of the corresponding developments in the real domain, a profound difference appears between functions of a real and functions of a complex argument after the introduction of differentiability: Whereas for a function $f(x)$ of a real variable, its differentiability need imply nothing at all concerning the possible higher derivatives,—as is well known, the first derivative $f'(x)$ need not be differentiable, or even continuous,—it turns out that for a function $f(z)$ of a complex variable, the existence of a first derivative automatically implies *the existence of all higher derivatives.* Formulated more precisely, there is the following

THEOREM. *If a function $f(z)$ is defined in a region $\mathfrak{G}$, and if it has a derivative $f'(z)$ there, then it also possesses all higher derivatives $f''(z)$, $f'''(z)$, ... in $\mathfrak{G}$.* (By a *region* is meant an open and *connected* point-set, i.e., an open set of points such that every pair of its points can be joined by a segmental arc belonging entirely to this point set.)

We cannot prove this theorem here. It is rather deep, and can be proved only after further development of the theory of functions with the aid of its integral calculus.[66] It does, however, make it appear understandable why those functions which are differentiable in regions have been given a special name:

DEFINITION. *A function $f(z)$ which is differentiable in a region $\mathfrak{G}$ is called a regular analytic (or also merely: a regular, or simply: an analytic) function in $\mathfrak{G}$. The region $\mathfrak{G}$ is called a region of regularity of the function. At every single point of $\mathfrak{G}$, the function is said to be regular.*

[66]Cf. *Th. F. I*, §16.

A rational function is regular in the entire plane, from which the zeros of its denominator have been deleted. Every power series represents, in its circle of convergence, an analytic function; this circle is a region of regularity of the represented function.

37. Conformal mapping

The behavior of a real function $y = f(x)$ of a real variable can be visualized in the familiar manner by means of its geometric image in an xy-plane. In the case of a function $w = f(z)$ of a complex argument, something analogous is not immediately possible, because each of the variables, z and w, has two coordinates. We get away from this difficulty by using *two* planes, a z-plane and a w-plane. In the first we plot the point z, in the second, the point $w = f(z)$ which the function makes correspond to z.[67] In this way, an *image point w* is associated with every point of $\mathfrak{M}$, the domain of definition of $f(z)$; in short: *The domain $\mathfrak{M}$ is mapped on the w-plane.* We are already familiar with this mapping in the case of the linear functions (see section II). We shall now determine for arbitrary functions $w = f(z)$, what corresponds, in the mapping, to the properties of continuity and differentiability.

The *continuity* of a function $w = f(z)$ at a point ζ is very easy to interpret geometrically. The second form of the definition given in §33 obviously asserts the following: If an (arbitrarily small) circle with radius $\epsilon > 0$ is described about the image point, $\omega = f(\zeta)$, of ζ, then it is always possible to draw such a small circle (call its radius δ) about the point ζ, that the images of *all* points in the interior of this circle about ζ lie within the circle chosen about ω. Thus, the image w lies in a *prescribed* neighborhood of ω, provided that the original point z lies in a *sufficiently small* neighborhood of ζ. In this sense (but also only in this sense) we may say briefly: Neighboring points in the z-plane are mapped into neighboring points in the w-

[67]Or, we imagine that to the point z in the z-plane, the corresponding functional value $w = f(z)$ is "attached", and that the point z is the *"bearer"* of the functional value w.

plane; or: To a small movement of z corresponds also a small movement of the image w.

From this follows, in particular: If $f(z)$ is continuous at every point of a region, then the image of every continuous curve in the region is again a continuous curve.[68]

Somewhat less simple, but of fundamental importance, is the *geometric interpretation of differentiability*. We obtain it in the following manner: Let $w = f(z)$ be defined in a circular region $\Re$ with center ζ, and suppose that $f(z)$ is differentiable at ζ. *Let the derivative $f'(\zeta)$ be different from* 0. We shall further *assume* that two distinct points z in $\Re$ yield also two distinct image points w,[69] and shall confine the rest of our considerations to $\Re$. Now, let $\mathfrak{k}$ be an arbitrary (oriented) arc issuing from ζ and possessing a (half-) tangent t at ζ. Then we show first:

The image curve $\mathfrak{k}'$ also has a tangent, t', at the image point $\omega = f(\zeta)$, and the direction of t' is that of t rotated through the angle am $f'(\zeta)$ in the positive sense.

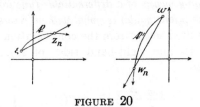

FIGURE 20

Proof: On $\mathfrak{k}'$ we choose a sequence of points $\{w_n\}$, whose terms are all different from ω, and such that $w_n \to \omega$ (Fig. 20). Let z_n be the original point corresponding to w_n. Then the sequence $\{z_n\}$ lies on $\mathfrak{k}$, its terms are all different from ζ, and $z_n \to \zeta$. Hence (see §34, 3),

$$(1) \qquad \frac{w_n - \omega}{z_n - \zeta} \to f'(\zeta),$$

[68]In particular cases, this curve may degenerate; e.g., if $f(z)$ is identically constant.

[69]In the further development of the theory of functions, it is shown that this is automatically the case under the hypothesis $f'(\zeta) \neq 0$, if the radius of $\Re$ is not too large, and if $f(z)$ is regular at ζ. (Cf. *Th. F. I*, §34.)

and, in particular (for suitable determination of the angles that appear),

$$\text{(2)} \qquad \text{am } (w_n - \omega) - \text{am } (z_n - \zeta) \to \text{am } f'(\zeta).$$

The two angles on the left-hand side are the vectorial angles of the secants from ω to w_n, ζ to z_n, respectively. Since f is supposed to have a tangent at ζ, am $(z_n - \zeta)$ tends to the vectorial angle of this tangent t as $n \to \infty$. Call this angle τ, and set am $f'(\zeta) = \alpha$. Then, according to (2),

$$\text{(3)} \qquad \text{am } (w_n - \omega) \to \tau + \alpha.$$

But this means precisely that f' also has a tangent and that its vectorial angle is equal to $\tau + \alpha$, Q.E.D.

If we allow two curves to emanate from ζ, forming the angle γ, then it follows from what precedes, that the image curves also form the angle γ (because the mapping turns both tangents through the *same* angle α); i.e.,

The mapping by means of a differentiable function $w = f(z)$ is (under the assumptions made) isogonal without reversion of angles.

This fact resulted solely from the consideration of the amplitudes of the left- and right-hand sides of (1). That, correspondingly, also

$$\text{(4)} \qquad \frac{|\, w_n - \omega \,|}{|\, z_n - \zeta \,|} \to |\, f'(\zeta) \,|,$$

likewise expresses an important geometric property of the mapping. Here on the left, in numerator and denominator, are the *lengths* of the vectors from ω to w_n, ζ to z_n, respectively. If we call them, briefly, *corresponding vectors*, then (4) asserts that the lengths of corresponding vectors issuing from ζ and ω are approximately in the ratio $1 : |\, f'(z) \,|$, provided that both lengths are sufficiently small. We express this, somewhat loosely, as follows: All "infinitely small" vectors issuing from ζ are stretched in the same ratio $1 : |\, f'(\zeta) \,|$ (which depends only on ζ). The mapping is therefore said to *preserve scale* ("in the infinitely small").

A mapping which, in the sense explained, is isogonal and, at the same time, preserves scale, is said to be *conformal*.

Summing up, we can therefore say briefly:

The mapping effected by an analytic function is conformal in a neighborhood of every point at which the derivative of the function does not vanish.

If we have three points z lying close to one another, then, according to what has preceded, the three image points w form a triangle which is *nearly* similar to that formed by the three points z,—and, indeed, the smaller the triangles, the greater the similarity. The mapping described is therefore also said to be a *similarity in the smallest parts*.

A linear function is regular in the entire plane with the exception of at most one point, and, according to §34, Example 4, its derivative is nowhere equal to 0. Hence, the preceding discussion affords a new proof of the isogonality of the mapping effected by a linear function, and we see, moreover, that this mapping is conformal. We shall become acquainted, in the next section, with additional conformal mappings.

CHAPTER XI

POWER AND ROOT.
THE RATIONAL FUNCTIONS

38. Power and root

The simplest of the rational functions, after the linear functions which we became acquainted with in section II, are the powers, i.e., the functions

$$(1) \qquad w = z^k,$$

where k denotes a natural number which we shall immediately think of as being greater than 1. We already know that such a function is continuous and differentiable, and hence analytic, in the entire plane, and that its derivative is $w' = kz^{k-1}$. Consequently, the derivative is different from 0 in the entire plane except at the origin. The mapping of the z-plane effected by the function (1) is therefore conformal everywhere except at the origin.

We shall investigate this mapping for the case $k = 2$, i.e., the mapping by means of the function

$$(2) \qquad w = z^2,$$

somewhat more closely. We show first:

By means of (2), the (open) right half-plane is mapped, in a one-to-one and, without exception, conformal manner, on the w-plane which has been cut along the negative axis of reals; i.e., on the totality of points in the w-plane which are different from 0 and the negative real points.

For if we set, as heretofore, $|z| = \rho$ and am $z = \varphi$, then,

106

according to §11 (5),

(3) $$|w| = \rho^2, \qquad \text{am } w = 2\varphi.$$

Now, if z describes the semicircle $|z| = \rho$, $-\pi/2 < \varphi < +\pi/2$ which lies in the right half-plane, w describes the *full* circle $|w| = \rho^2$, $-\pi < \text{am } w < +\pi$ which has been cut at the point

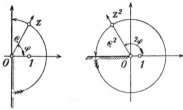

FIGURE 21

-1, and both curves correspond in a one-to-one manner (Fig. 21). If we let ρ run through all values in $0 < \rho < +\infty$, ρ^2 also runs through all these values, and assumes each precisely once. Herewith the assertion is already proved, because the derivative of our function (2) vanishes nowhere in $\Re(z) > 0$. The one-to-one character of the mapping is retained if the positive axis of imaginaries is added to the right half-plane $\Re(z) > 0$, and if the "upper" boundary of the cut w-plane is added to the latter. The isogonality, however, is destroyed at the origin, because, according to (3), the angles at the origin are doubled under the mapping.

In the same manner, we see that the *left* half-plane $\Re(z) < 0$, too, to which the *negative* axis of imaginaries has been added, (for whose points, consequently, $+\pi/2 < \varphi \leqq 3\pi/2$), is likewise mapped by (2), in a one-to-one manner, on the w-plane which has been cut and bounded exactly as before, and that the mapping is conformal except at the origin.[70] The *full* z-plane is thus mapped, in a clearly perceived way, on the doubly covered w-plane; i.e., to every z corresponds precisely one w, but every w is obtained for precisely two values of z (which differ only in sign),—with the exception of the value $w = 0$,

[70]This, of course, also follows directly from what was proved previously, because $(-z)^2 = z^2$.

which is assumed only for $z = 0$. For the purpose of visualizing more clearly this double covering of the w-plane, it is useful to imagine one of the two previously obtained copies of the cut w-plane to be placed *upon the other*. If we then fasten the two origins together, and join the sheets "crosswise," i.e., fuse the upper boundary of each sheet with the lower boundary of the other,[71] we obtain a peculiar configuration, which is called a *Riemann surface*. On it, every point different from 0 appears *twice* (at superposed positions), the origin, however, only precisely *once*. Our function $w = z^2$ now maps the (simple[72]) z-plane on this Riemann surface in a one-to-one and also, apart from the *winding point* or *branch-point* at 0, conformal manner.— However, we cannot enter here into a more detailed treatment of such Riemann surfaces.[73]

We also obtain a good insight into the mapping effected by (2), by using Cartesian coordinates. Let us set $z = x + iy$, $w = u + iv$. Then, according to (2),

$$(4) \qquad u = x^2 - y^2, \qquad v = 2xy.$$

From this we infer that the points z lying on the hyperbolas $x^2 - y^2 = $ const. go over into the points on the lines $u = $ const. Likewise, the hyperbolas $xy = $ const. go over into the straight lines $v = $ const. Because of the isogonality of the mapping, every hyperbola of the one family intersects every one of the other family at right angles.

It is equally easy to see, from (4), that the two families of straight lines $x = $ const., $y = $ const. are mapped into two confocal families of parabolas, with focus at 0, which again are orthogonal to each other.

With the use of polar coordinates, the mapping effected by the function (1) with $k > 2$ is just as easy to study as the case $k = 2$. We have merely to replace the half-plane by an angular region with vertex at 0 and an aperture of $2\pi/k$ radians; and the double covering of the w-plane becomes a k-fold one. Even

[71]This can be carried out only in imagination, since the penetration of the two sheets of a material model is merely imperfectly realizable.

[72]In German: *schlicht*.

[73]For a discussion of these surfaces, see *Th. F. II*, section II.

when Cartesian coordinates are employed, no difficulties in principle arise. The curves, for $k > 2$, corresponding to the hyperbolas and parabolas, are merely not so simple any more; they are algebraic curves of the kth order.

Since the mapping between the simple z-plane and the k-tuply covered w-plane is (if we disregard the origin) one-to-one, we can, without further ado, interchange z and w in the entire discussion; i.e., we can regard w as the given value and z as the value associated with w. We then immediately obtain:

For a given $w \neq 0$, there exist precisely k distinct values z for which $z^k = w$. These values all lie on the same circle about the origin in the z-plane, and constitute there the vertices of a regular k-gon.

Each of these values is called a *kth root of w*; in symbols: $z = \sqrt[k]{w}$. This symbol is thus—in antithesis to the usual conventions made in the real domain,—by nature, a *multiple-valued*, namely, a *k-valued, symbol.*

This can be realized, independently of what precedes, as follows: If $w = \sigma (\cos \psi + i \sin \psi)$, $z = \rho (\cos \varphi + i \sin \varphi)$, then, because of (1), we must have

$$(5) \qquad \rho^k = \sigma, \qquad k\varphi = \psi.$$

Now, ρ and σ are positive. Therefore, the first of these equations is satisfied, for given σ, by *precisely one* value ρ, namely, by the root $\rho = \sqrt[k]{\sigma}$ (which, in the real domain, is *uniquely* determined and again positive). However, since the equality of two angles merely signifies their congruence mod 2π, the second equation is satisfied by k distinct angles φ, namely, in addition to the value $\varphi = (1/k)\psi$, the values $(1/k)(\psi + 2\pi)$, $(1/k)(\psi + 4\pi)$, ..., $(1/k)(\psi + 2(k - 1)\pi)$, and only these. Hence, $\sqrt[k]{w}$, for $w \neq 0$, has the k values

$$
\begin{aligned}
&\sqrt[k]{\sigma} \left(\cos \frac{\psi + 2\nu\pi}{k} + i \sin \frac{\psi + 2\nu\pi}{k} \right), \\
&(6) \qquad\qquad\qquad \nu = 0, 1, 2, \cdots, k - 1.
\end{aligned}
$$

$\sqrt[k]{0}$, on the other hand, is to be set equal to the sole value 0. As the *principal value* of $\sqrt[k]{w}$ we designate that one of the values (6) which is obtained by taking for ψ the principal value of am w and setting $\nu = 0$.—We cannot enter further, here, into the study of the root functions.

39. *The entire rational functions*

The close investigations of the *entire rational functions* or, as we say briefly, *polynomials*, that is, the functions of the form

$$(1) \qquad w = a_0 + a_1 z + a_2 z^2 + \cdots + a_p z^p,$$

is the principal subject of classical algebra, into a discussion of which we shall, of course, not enter here. We should, however, at least mention the theorem which occupies a particularly important position in it, and which has, for that reason, been named the *fundamental theorem of algebra*. It has been (cf. §4), above all, the possibility of proving this theorem, that has prepared the way for the universal recognition of the complex numbers.

FUNDAMENTAL THEOREM OF ALGEBRA. *Every polynomial in z,*

$$(2) \qquad g(z) = a_0 + a_1 z + \cdots + a_p z^p \qquad (a_p \neq 0),$$

whose "degree" $p \geqq 1$, can be decomposed into precisely p linear factors; i.e., there exist p (not necessarily distinct) numbers, $z_1, z_2, \ldots, z_p$, such that

$$(3) \qquad g(z) = a_p(z - z_1)(z - z_2) \ldots (z - z_p).$$

In addition to the proofs given in algebra, there are several purely function-theoretical proofs, two of which are to be found in *Th. F. I*, §§28 and 35.

The distinct numbers among $z_1, z_2, \ldots, z_p$ are called, briefly, the *roots* or *zeros* of the polynomial (2), and we shall denote them by $\zeta_1, \zeta_2, \ldots, \zeta_k$, $(1 \leqq k \leqq p)$. If ζ_ν appears among $z_1, \ldots, z_p$ a total of α_ν times, $(\nu = 1, 2, \ldots, k)$, we say that ζ_ν is a root or zero of *order* α_ν. Naturally, $\alpha_1 + \alpha_2 + \cdots + \alpha_k = p$, and instead of (3) we can write

$$(4) \qquad g(z) = a_p(z - \zeta_1)^{\alpha_1}(z - \zeta_2)^{\alpha_2} \ldots (z - \zeta_k)^{\alpha_k}.$$

This representation of a polynomial is called its *factor representation*.

40. *The fractional rational functions*

A rational function $f(z)$ is said to be *fractional*, if it cannot be represented as an entire rational function. $f(z)$ can then be brought into the form

(1)
$$f(z) = \frac{h(z)}{g(z)},$$

where $h(z)$ and $g(z)$ are polynomials. Concerning these fractional rational functions, too, we shall speak of only one theorem here, the theorem *on the partial-fractions decomposition of rational functions*, which is usually proved in algebra, but which can be proved in the theory of functions only after further development of the latter (see *Th. F. I*, §35). A rational function of the particularly simple form

$$\frac{c}{(z - \zeta)^\gamma},$$

where both $c \neq 0$ and ζ denote complex numbers and γ denotes a natural number, is called a *partial fraction*. With this designation we have the following

THEOREM. *Every rational function can be represented—and, essentially, only in precisely one way—as the sum of an entire rational function and a finite number of partial fractions.*

More precisely: If we are concerned with the rational function (1), and if its denominator $g(z)$ has the factor representaton given in §39 (4), then there is exactly one polynomial $q(z)$, and there are exactly p complex numbers $c_{\kappa\lambda}$, such that $f(z)$ possesses the representation

$$f(z) = q(z) + \frac{c_{11}}{z - \zeta_1} + \frac{c_{12}}{(z - \zeta_1)^2} + \cdots + \frac{c_{1\alpha_1}}{(z - \zeta_1)^{\alpha_1}}$$

$$+ \frac{c_{21}}{z - \zeta_2} + \frac{c_{22}}{(z - \zeta_2)^2} + \cdots + \frac{c_{2\alpha_2}}{(z - \zeta_2)^{\alpha_2}}$$

$$+ \cdots\cdots\cdots\cdots\cdots\cdots\cdots\cdots\cdots\cdots\cdots$$

$$+ \frac{c_{k1}}{z - \zeta_k} + \frac{c_{k2}}{(z - \zeta_k)^2} + \cdots + \frac{c_{k\alpha_k}}{(z - \zeta_k)^{\alpha_k}}.$$

Simple assertions concerning the conformal mapping effected by rational functions which differ from the linear functions, can be made only in special cases.

CHAPTER XII

THE EXPONENTIAL, TRIGONOMETRIC, AND HYPERBOLIC FUNCTIONS

41. The exponential function

The power series

$$(1) \qquad 1 + z + \frac{z^2}{2!} + \cdots + \frac{z^n}{n!} + \cdots = \sum_{n=0}^{\infty} \frac{z^n}{n!}$$

is, as was already established in §29, 3, everywhere convergent. It therefore defines a regular analytic function in the entire plane, or, as we say briefly, an *entire function*. It is well known from the differential and integral calculus, that, for real $z = x$, the series (1) represents the exponential function e^x. The function of a complex argument represented by (1) in the entire z-plane is therefore also called the *exponential function*, and is denoted by e^z. The power e^z is thus *defined* for complex exponents z in a *single-valued* manner by means of the relation

$$(2) \qquad e^z = \sum_{n=0}^{\infty} \frac{z^n}{n!}.$$

We are entitled to make this definition, first, because the symbol e^z, for complex z, has had no meaning at all up to now, and further, because the meaning now laid down proves to be useful and significant, as subsequent investigation will show. Moreover, according to §35, Theorem 2, there can exist, in addition to (1), no other power series with 0 as center, which has, for the real values of $z = x$ lying in a neighborhood of 0, the same sum e^x as (1). Our series *continues*, as we say, *the real function e^x into the complex domain*. The following facts will show that the properties of the real function e^x also belong, in large measure, to the analytic function e^z; we shall, however, become acquainted with important new properties of e^z.

1. In the complex domain, as in the real domain, we have

113

the addition-theorem

$$(3) \qquad e^{z_1} \cdot e^{z_2} = e^{z_1 + z_2}.$$

For since the series (1) is absolutely convergent for every z, the series for e^{z_1} and e^{z_2} may be multiplied according to the Cauchy rule (see §28). The nth term of the product series then becomes

$$\frac{z_1^n}{n!} + \frac{z_1^{n-1}}{(n-1)!} \cdot \frac{z_2}{1!} + \cdots + \frac{z_1^{n-\nu}}{(n-\nu)!} \cdot \frac{z_2^\nu}{\nu!} + \cdots + \frac{z_2^n}{n!}$$

$$= \frac{1}{n!} \left[z_1^n + \binom{n}{1} z_1^{n-1} z_2 + \cdots + \binom{n}{\nu} z_1^{n-\nu} z_2^\nu + \cdots + z_2^n \right]$$

$$= \frac{(z_1 + z_2)^n}{n!}.$$

Consequently, we have indeed

$$\sum_{n=0}^{\infty} \frac{z_1^n}{n!} \cdot \sum_{n=0}^{\infty} \frac{z_2^n}{n!} = \sum_{n=0}^{\infty} \frac{(z_1 + z_2)^n}{n!},$$

which proves the assertion (3).

2. Formula (3) now enables us actually to calculate the value of e^z for given z. For if $z = x + iy$, then, according to (3),

$$e^z = e^{x+iy} = e^x \cdot e^{iy}.$$

By (2) we have, further,

$$e^{iy} = \sum_{n=0}^{\infty} \frac{(iy)^n}{n!} = \sum_{k=0}^{\infty} (-1)^k \frac{y^{2k}}{(2k)!} + i \sum_{k=0}^{\infty} (-1)^k \frac{y^{2k+1}}{(2k+1)!}.$$

Here on the right-hand side are two *real* power series whose sums are known, from real analysis, to be $\cos y$, $\sin y$, respectively. These values, as well as e^x, can be read off, for given x and y, from the ordinary logarithmic-trigonometric tables, and can therefore be regarded as known. Hence, the formula

$$(4) \qquad e^z = e^{x+iy} = e^x(\cos y + i \sin y)$$

renders possible the numerical calculation of e^z in a simple manner.

3. We read off from (4), that

$$(5) \qquad |e^z| = e^{\Re(z)} \qquad \text{and} \qquad \text{am } (e^z) = \Im(z),$$

and further, that the direction factor (see §11) of a complex number can now be written in the simpler form

(6) $$\cos \varphi + i \sin \varphi = e^{i\varphi}.$$

According to this, we have, in particular,

(7) $$e^{2\pi i} = 1,$$
$$e^{\pi i} = e^{-\pi i} = -1, \qquad e^{\pi i/2} = i, \qquad e^{-\pi i/2} = -i.$$

Furthermore, we see that the functional value e^z is *real* if, and only if, $\sin y = 0$, and hence $y = \Im(z) = k\pi$,—in other words, if, and only if, z lies on the hereby specified family of parallels to the axis of reals (or on this axis itself).

4. Formula (7), in connection with the addition-theorem (3), now shows that, for every z,

$$e^{z+2\pi i} = e^z \cdot e^{2\pi i} = e^z.$$

Thus, the exponential function is *periodic* with the period $2\pi i$: Its value does not change if the variable z is increased by $2\pi i$. Of course, we now have more generally, for every integral $k \gtrless 0$,

(8) $$e^{z+2k\pi i} = e^z.$$

At points of the z-plane which result from each other by means of a single or repeated application of the translation $(2\pi i)$ (or $(-2\pi i)$), e^z has the same value.

5. The converse is also true: If the equation $e^{z_1} = e^{z_2}$ is valid for two points z_1 and z_2, then they differ only by an integral multiple of $2\pi i$; i.e., we must have

(9) $$z_2 = z_1 + 2k\pi i.$$

For from $e^{z_2} = e^{z_1}$ it follows, first (according to (3)), that $e^{z_2-z_1} = 1$. Now, if

$$e^z = e^{x+iy} = 1,$$

then, by (4) and (5), we must have $e^x = 1$ and, at the same time, $\cos y = 1$, $\sin y = 0$. But this is the case only for $x = 0$, $y = 2k\pi$. Hence, it is necessary that

(9') $$z_2 - z_1 = 2k\pi i.$$

Every value w which is at all assumed by the exponential function $w = e^z$ is thus already assumed in a parallel strip, one of whose boundaries can be obtained from the other by means of the translation $(2\pi i)$. The strip

$$(10) \qquad -\pi < \Im(z) \leqq +\pi$$

is usually chosen as such a *fundamental domain* of the function e^z; the upper boundary is included, the lower one is not.

6. *In the fundamental strip* (10), *the exponential function assumes every non-zero value w precisely once. The value* 0, *however, is assumed nowhere.*

The last is almost self-evident; for, according to (3),

$$e^z \cdot e^{-z} = e^0 = 1,$$

and hence (by §8, Theorem), e^z cannot equal 0. Now, let $w \neq 0$, and set, as before, $|w| = \sigma$, am $w = \psi$. Then, we see immediately from (4), that, for

$$(11) \qquad z = \log \sigma + i(\psi + 2k\pi), \qquad (k \gtreqless 0, \text{ integral}),$$

obviously $e^z = w$. Precisely one of the values (11) lies in the fundamental strip (10). For any other value z' of this strip, however, $e^{z'}$ cannot also be equal to w, because of the result established in 5.

7. If we differentiate the series (1) term by term, we again obtain this series. Hence, for every z,

$$\frac{de^z}{dz} = e^z.$$

In particular, the derivative is, consequently, everywhere different from 0, and hence the entire z-plane is mapped conformally, without exception, by means of the function $w = e^z$.

8. This mapping is also easy to see in detail: In the fundamental strip (10) of Fig. 22, we imagine the straight lines parallel to the boundaries to be drawn, and to be oriented from left to right; likewise, we imagine the segments perpendicular to these lines to be drawn from boundary to boundary, and to be oriented from bottom to top. If z describes one of the first-

named lines from left to right, this means that in $z = x + iy$ we hold the imaginary part y fixed and let x run through all real numbers in increasing order. According to (5), then, the image point w has the fixed amplitude y, and, consequently, lies on the ray which emanates from 0 and corresponds to this amplitude; while its absolute value e^x runs through the values from 0 to $+\infty$ in increasing order. The image point thus describes the ray in question from 0 (excl.) to ∞ (excl.): directed line and ray correspond in a one-to-one manner. If z describes one of the above-mentioned vertical segments from bottom to top, this means that we leave x fixed and let y run through the values from $-\pi$ (excl.) to $+\pi$ (incl.). According to (5), w, then, has the fixed absolute value e^x, and consequently, describes the circle with radius e^x about the origin in the w-plane precisely once in the positive sense, beginning at the negative axis of reals (excl.) and returning to it (incl.). Thus, in particular, the interior of the fundamental strip is mapped in a one-to-one and, without exception, conformal manner on the interior of the w-plane cut along the negative axis of reals.

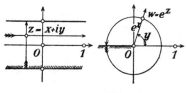

FIGURE 22

9. Of particular interest for various investigations (see §43) is the problem of developing the function

$$(12) \qquad w = \frac{z}{e^z - 1} = \frac{1}{1 + \dfrac{z}{2!} + \dfrac{z^2}{3!} + \cdots}$$

in a power series with 0 as center. According to §30, this is possible, at all events, for a certain neighborhood of the origin. The coefficients of the expansion obtained are denoted, for historical reasons, by $B_n/n!$, so that we set

$$(13) \qquad \frac{1}{1 + \dfrac{z}{2!} + \dfrac{z^2}{3!} + \cdots} = 1 + B_1 z + \frac{B_2}{2!} z^2 + \cdots .$$

Now, instead of calculating the coefficients according to the general method of §30, we proceed here, and in all analogous cases, as follows: By (13),

$$\left(1 + \frac{z}{2!} + \frac{z^2}{3!} + \cdots \right)\left(1 + B_1 z + \frac{B_2}{2!} z^2 + \cdots \right) = 1;$$

i.e., on multiplying out the two power series on the left, we must obtain a power series whose constant term is equal to 1 and whose remaining coefficients are all equal to 0. This yields the infinitely many equations

$$\frac{1}{1!} \frac{B_n}{n!} + \frac{1}{2!} \frac{B_{n-1}}{(n-1)!} + \cdots + \frac{1}{n!} \frac{B_1}{1!} + \frac{1}{(n+1)!} = 0,$$

$$(n = 1, 2, \ldots).$$

After multiplication by $(n + 1)!$, the binomial coefficients of the $(n + 1)$-th power appear here on the left. The equations therefore read

$$(14) \qquad \begin{cases} 2B_1 + 1 & = 0 \\[4pt] 3B_2 + \ 3B_1 + 1 & = 0 \\[4pt] 4B_3 + \ 6B_2 + \ 4B_1 + 1 & = 0 \\[4pt] 5B_4 + 10B_3 + 10B_2 + 5B_1 + 1 = 0 \\[4pt] \cdots\cdots\cdots\cdots\cdots\cdots\cdots\cdots\cdots\cdots\cdots , \end{cases}$$

from which we obtain successively

$$B_1 = -\frac{1}{2}, \ B_2 = \frac{1}{6}, \ B_3 = 0, \ B_4 = -\frac{1}{30}, \ B_5 = 0, \ B_6 = \frac{1}{42}, \cdots .$$

These numbers are called the *Bernoullian numbers*. They are, as the above calculation shows, all *rational* numbers. Since their calculation offers no difficulties in principle, they may be regarded as "known." With the exception of B_1 , all B_n with odd index n have the value 0. This follows (with the application of Theorem 2 in §35) from the fact that, as is easily verified,

$$\frac{z}{e^z - 1} + \frac{z}{2}$$

is an *even* function of z, i.e., has the same value for $-z$ as for z.

42. The functions cos z and sin z

Considerations entirely analogous to those appearing at the beginning of the preceding paragraph lead us, by necessity, to *define* the trigonometric functions cos x and sin x for complex variables by means of the relations

$$(1) \qquad \cos z = 1 - \frac{z^2}{2!} + \frac{z^4}{4!} - + \cdots = \sum_{k=0}^{\infty} (-1)^k \frac{z^{2k}}{(2k)!}$$

and

$$(2) \qquad \sin z = z - \frac{z^3}{3!} + \frac{z^5}{5!} - + \cdots = \sum_{k=0}^{\infty} (-1)^k \frac{z^{2k+1}}{(2k + 1)!}.$$

Since the two series, just as the exponential series, converge everywhere, cos z and sin z are also hereby defined as *entire functions*. cos z is an *even* function, sin z is an *odd* function; i.e., for every z,

$$(3) \qquad \cos(-z) = \cos z, \qquad \sin(-z) = -\sin z.$$

The connection between our three series, which was already employed for real variables in §41, 3, now obviously also sub- ·sists for complex variables; i.e., for every complex z we have what are known as *Euler's formulas*:

$$(4) \qquad\qquad e^{iz} = \cos z + i \sin z,$$

$$(5) \qquad \cos z = \frac{e^{iz} + e^{-iz}}{2}, \qquad \sin z = \frac{e^{iz} - e^{-iz}}{2i}.$$

To prove these, we need only substitute for the occurring functional values, the power series defining them, whereupon, on both sides of each of the equations, the same series is obtained.

Because of this extremely simple connection between cos z and sin z on the one hand and e^z on the other, the investigations of cos z and sin z present no new difficulties. Everything follows very simply from the facts ascertained in §41.

1. *The addition-theorems for the functions* cos *and* sin, *known from the real domain, are also valid in the complex domain*; i.e., for arbitrary complex numbers z_1 and z_2 we have always

(6)
$$\cos (z_1 + z_2) = \cos z_1 \cos z_2 - \sin z_1 \sin z_2 ,$$

$$\sin (z_1 + z_2) = \cos z_1 \sin z_2 + \sin z_1 \cos z_2 .$$

For according to (5) and §41 (3),

$$\cos (z_1 + z_2) = \frac{e^{iz_1}e^{iz_2} + e^{-iz_1}e^{-iz_2}}{2},$$

from which, with the use of (4), the first of the formulas (6) immediately follows. The second is proved in an entirely analogous manner.

2. *The periodicity properties, known from the real domain, are also retained in the complex domain. Both functions have the (real) period* 2π; *i.e., for every* z,

$$\cos (z + 2\pi) = \cos z, \qquad \sin (z + 2\pi) = \sin z.$$

To prove this, we have only to apply the addition-theorems just established, to the expressions on the left-hand sides, noting that cos $2\pi = 1$, sin $2\pi = 0$.

3. Since the facts established in 1. and 2. are formally the same as in the real domain, all consequences which can be deduced purely formally from these facts also continue to hold. *But this is the entire formula apparatus of goniometry,* as it is called. Thus, e.g., the formulas

$$\cos^2 z + \sin^2 z = 1,$$

$$\cos 2z = \cos^2 z - \sin^2 z, \qquad \sin 2z = 2 \cos z \sin z,$$

$$\cos z_1 + \cos z_2 = 2 \cos \frac{z_1 + z_2}{2} \cos \frac{z_1 - z_2}{2}, \text{ etc.,}$$

are valid, without change, for arbitrary complex arguments z, z_1, z_2. It is not necessary to write down all these formulas in detail.

4. The *calculation* of the functional values $\cos z$ and $\sin z$ does not raise any difficulties either. For, making use of (5) and (4), we have, for $z = x + iy$,

$$\cos (x + iy) = \frac{1}{2} (e^{ix-y} + e^{-ix+y})$$

(7)

$$= \cos x \cdot \frac{e^y + e^{-y}}{2} - i \sin x \cdot \frac{e^y - e^{-y}}{2},$$

and similarly

$$(8) \qquad \sin (x + iy) = \sin x \cdot \frac{e^y + e^{-y}}{2} + i \cos x \cdot \frac{e^y - e^{-y}}{2}.$$

5. *The zeros of* $\cos z$ *and* $\sin z$, *which are known to us from the real domain, are also the only ones in the complex domain.* For if we are to have $\cos z = 0$, then, according to (5), we must have $e^{iz} = -e^{-iz}$ or $e^{2iz} = -1 = e^{\pi i}$. Hence, by §41 (9), it is necessary that $2iz = \pi i + 2k\pi i$, i.e.,

$$z = (2k + 1) \frac{\pi}{2}.$$

And if we are to have $\sin z = 0$, it follows, analogously, that we must have $z = k\pi$, ($k \gtreqless 0$, integral).

6. $\cos z_2 = \cos z_1$ *if, and only if,* $z_2 = \pm z_1 + 2k\pi$,—i.e., under the same conditions as in the real domain. For since

$$\cos z_2 - \cos z_1 = 2 \sin \frac{z_1 + z_2}{2} \sin \frac{z_1 - z_2}{2},$$

this difference can be equal to zero only if (see 5.) either $(z_1 + z_2)/2$ or $(z_1 - z_2)/2$ is an integral multiple of π. In a similar manner it follows that $\sin z_2 = \sin z_1$ *if, and only if,*

$$z_2 = z_1 + 2k\pi \qquad or \qquad z_2 = \pi - z_1 + 2k\pi,—$$

again as in the real domain.

7. *The functions* cos z *and* sin z *assume, in a period-strip, say in the strip*

(9) $$-\pi < \Re(z) \leqq +\pi,$$

every value distinct from ± 1, *at precisely two points, whereas each of the two values* ± 1 *is assumed at precisely one point of the strip.*

At the point 0, cos $z = 1$; at the point π, cos $z = -1$. According to 6., these values can be assumed at no second point of the strip. However, if w is an arbitrary complex number distinct from ± 1, and if we are to have cos $z = w$, then, by (5), we must have

$$e^{iz} + e^{-iz} = 2w \quad \text{and hence} \quad e^{iz} = w + \sqrt{w^2 - 1}.$$

Since $w \neq \pm 1$, the symbol $\sqrt{w^2 - 1}$ has precisely two distinct values, and hence the same is true for $w + \sqrt{w^2 - 1}$, whose two values we call w_1 and w_2 According to §41, 6., each of the two equations $e^z = w_1$ and $e^z = w_2$ has precisely one solution z in the strip $-\pi < \Im(z) \leqq +\pi$. Consequently, each of the equations

$$e^{iz} = w_1, \qquad e^{iz} = w_2$$

has precisely one solution z in the strip $-\pi < \Re(z) \leqq +\pi$. If we call the solutions z_1, z_2, respectively, then $z_1 \neq z_2$ and cos $z_1 = $ cos $z_2 = w$. By 6., however, there can be no third point z at which cos $z = w$.—For the function sin z, the proof is entirely analogous.

8. By differentiating the series (1) and (2) term by term, it follows, as in the real domain, that

$$\frac{d \cos z}{dz} = -\sin z, \qquad \frac{d \sin z}{dz} = \cos z$$

for every z. Thus, the mapping effected by the function $w = $ cos z is conformal everywhere except at the points $k\pi$; the mapping effected by $w = $ sin z, everywhere except at the points $(2k + 1)\pi/2$.

9. The details of these mappings follow from formulas (7) and (8). (7) shows, e.g., that under the mapping effected by

$w = \cos z$, the straight lines (segments) which are parallel (perpendicular) to the boundaries of the period-strip, go over into confocal hyperbolas (ellipses). We must leave it to the reader to carry out the (very simple) proof.

43. The Functions tan z and cot z

The functions $\tan z$ and $\cot z$ are defined for complex variables, as in the real domain, by means of the relations

$$(1) \qquad \tan z = \frac{\sin z}{\cos z}, \qquad \cot z = \frac{\cos z}{\sin z}.$$

Since $\cos z$ and $\sin z$ are regular in the entire plane, so is $\tan z$ at all points where $\cos z \neq 0$, that is, in the entire plane with the exception of the points $(2k + 1)\pi/2$. Similarly, $\cot z$ is regular in the entire plane, except at the points $k\pi$. The further properties can be derived easily from the properties of the functions $\cos z$, $\sin z$, and e^z, now familiar to us:

1. The power-series expansion of each of the two functions follows from the division problem treated in §41, 9. We have

$$(2) \quad \cot z = \frac{\cos z}{\sin z} = i\,\frac{e^{iz} + e^{-iz}}{e^{iz} - e^{-iz}} = i\,\frac{e^{2iz} + 1}{e^{2iz} - 1} = i + \frac{2i}{e^{2iz} - 1},$$

and hence, by §41, 9,

$$z \cot z = iz + \frac{2iz}{e^{2iz} - 1} = iz + 1 + B_1(2iz) + \frac{B_2}{2!}\,(2iz)^2 + \cdots.$$

Now, since $B_1 = -1/2$, and since all Bernoullian numbers with odd index vanish,

$$
\begin{aligned}
z \cot z &= 1 - \frac{2^2 B_2}{2!}\,z^2 + - \cdots + (-1)^k\,\frac{2^{2k}B_{2k}}{(2k)!}\,z^{2k} + \cdots \\[2mm]
&= 1 - \frac{1}{3}\,z^2 - \frac{1}{45}\,z^4 - \frac{2}{945}\,z^6 - \cdots .
\end{aligned}
$$

(3)

Since, now, further, $\tan z = \cot z - 2 \cot 2z$, the expansion of $\tan z$ follows immediately from (2):

$$\tan z = \frac{4 \cdot 3}{2!} B_2 z + \cdots$$

(4)
$$+ (-1)^{k-1} \frac{2^{2k}(2^{2k} - 1)}{(2k)!} B_{2k} z^{2k-1} + \cdots$$

$$= z + \frac{1}{3} z^3 + \frac{2}{15} z^5 + \frac{17}{315} z^7 + \cdots.$$

Each of these power series has certainly a positive radius of convergence. Its exact value, however, can be found only after deeper considerations; it is π for the series (3), $\pi/2$ for the series (4). (Cf. *Th. F. I*, §31.)

2. The formal properties, which find their expression in the addition-theorems and the remaining goniometric formulas (e.g., in the one just used to prove (4)), are, naturally, for the same reason as in the case of cos z and sin z, the same as in the real domain. We may therefore forego writing down these formulas in detail.

3. From the addition-theorem, it follows that the periodicity properties, too, are the same as in the real domain; for every z,

(5) $\tan (z + \pi) = \tan z, \qquad \cot (z + \pi) = \cot z.$

It is customary to choose the strip

$$-\frac{\pi}{2} < \Re(z) \leqq +\frac{\pi}{2}$$

as *period-strip.*—Similar to the case of cos z and sin z, it can also be shown here, more precisely: If $\tan z_1 = \tan z_2$, then z_1 and z_2 differ only by an integral multiple of π, and the same holds for cot z. Thus, *our functions possess the same value at two distinct points if, and only if, one of these points can be obtained from the other by means of a single or repeated application of the translation (π).* For from $\tan z_1 = \tan z_2$, as well as from $\cot z_1 = \cot z_2$, it follows that we must have $\sin (z_2 - z_1) = 0$, and hence (see §42, 5) $z_2 - z_1 = k\pi$.

4. *In the period-strip $-\pi/2 < \Re(z) \leqq + \pi/2$, tan z and cot z assume every complex value, distinct from $\pm i$, precisely once; the values $\pm i$, on the other hand, are not assumed at all.* It suffices

to prove this for cot z. For since $\tan z \cdot \cot z = 1$, the assertion concerning $\tan z$ then follows automatically. Now, the equation $\cot z = \pm i$ would mean, by (2), that $2i/(e^{2iz} - 1)$ would have to equal 0 or $-2i$. The first is certainly possible for no value of z, and the second would require that $e^{2iz} = 0$, which also is never the case. Hence, we have always $\cot z \neq \pm i$. If, however, w is an arbitrary number different from $\pm i$, then the equation $\cot z = w$ signifies, according to (2), that

$$(6) \qquad i\,\frac{e^{2iz} + 1}{e^{2iz} - 1} = w \qquad \text{or} \qquad e^{2iz} = \frac{w + i}{w - i}.$$

Since we have here on the right a definite and, moreover, non-zero value, there exists, by §41, 6, precisely one number z' in $-\pi < \Im(z') \leqq +\pi$, for which $e^{z'} = \dfrac{w + i}{w - i}$. Hence, there exists also only precisely one value z with $-\pi/2 < \Re(z) \leqq +\pi/2$, for which the second of the equations (6) holds, i.e., for which $\cot z = w$.

5. The derivatives of our functions result, of course, simply from the defining equations (1). We have, as in the real domain,

$$\frac{d\tan z}{dz} = \frac{1}{\cos^2 z}, \qquad \frac{d\cot z}{dz} = -\frac{1}{\sin^2 z}.$$

These derivatives obviously vanish nowhere. The mapping effected by either one of our functions is therefore conformal at every point at which the function is defined. We cannot go into details of these mappings.

44. *The hyperbolic functions*

For various applications it is useful to introduce, besides the trigonometric functions, the so-called *hyperbolic functions* cosh z, sinh z.[74] They are defined by means of the everywhere-convergent power-series

$$(1) \qquad \cosh z = 1 + \frac{z^2}{2!} + \frac{z^4}{4!} + \cdots,$$

---

[74]Sometimes, in addition, the functions $\tanh z = \sinh z/\cosh z$ and $\coth z = \cosh z/\sinh z$ are used; however, we shall not consider them here.

$$(2) \qquad \sinh z = z + \frac{z.}{3!} + \frac{z.}{5!} + \cdots ,$$

and are therefore *entire functions*. They are related to cos z and sin z through the simple formulas

$$(3) \qquad \cosh z = \cos (iz), \qquad \sinh z = -i \sin (iz),$$

as the series representations show. It follows from these formulas, that all properties of the new functions and the formula apparatus valid for them turn out to be very similar to those of cos z and sin z, so similar, that it is unnecessary to carry out all details. We therefore call attention, without proof, only to what is most important:

1. cosh z is an even function, sinh z is an odd function.

2. $\cosh z = \frac{1}{2}(e^z + e^{-z})$, $\sinh z = \frac{1}{2}(e^z - e^{-z})$, $e^z = \cosh z + \sinh z$.

3. The addition-theorems read:

$$\cosh (z_1 + z_2) = \cosh z_1 \cosh z_2 + \sinh z_1 \sinh z_2 ,$$

$$\sinh (z_1 + z_2) = \cosh z_1 \sinh z_2 + \sinh z_1 \cosh z_2 .$$

4. From these we get, e.g.,
$\cosh^2 z - \sinh^2 z = 1$, $\cosh 2z = \cosh^2 z + \sinh^2 z$, etc.

5. Both functions are periodic and have the period $2\pi i$.

6. $\quad \dfrac{d \cosh z}{dz} = \sinh z, \qquad \dfrac{d \sinh z}{dz} = \cosh z.$

CHAPTER XIII

THE LOGARITHM, THE CYCLOMETRIC FUNCTIONS, AND THE BINOMIAL SERIES

45. The logarithm

The natural logarithm is defined, as in the real domain, to be the inverse of the exponential function. But since the latter has turned out to be periodic in the complex domain, somewhat more profound differences from the real domain than heretofore, appear upon further investigation of the logarithm.

DEFINITION. *The number b shall be said to be a (not the) natural logarithm of a—in symbols*:

$$b = \log a -$$

if $e^b = a$.

Consequently, by §41, 5 and 6, every number different from 0 has *infinitely many* natural logarithms. Precisely one of these, it is called the *principal value* of the logarithm of a, is such that its imaginary part lies between $-\pi$ (excl.) and $+\pi$ (incl.). All remaining logarithms of the same number a differ from the principal value only by an additive multiple of $2\pi i$. If we denote the principal value by Log a, say, then the formula

$$(1) \qquad \log a = \text{Log } a + 2k\pi i, \qquad (k = 0, \pm 1, \pm 2, \ldots),$$

furnishes all the values of log a. No logarithm can be defined for the number 0, however, (because of §41, 6).

Moreover, if

(2) $|a| = A$, and the principal value of am a is α, then it follows immediately from §41, 5, that

$$(3) \qquad\qquad \text{Log } a = \log A + i\alpha,$$

if log A is understood to be the (unique, real) natural logarithm of the positive number A, known to us from real analysis. Thus, all values of log a have the same real part, log A, and the imaginary parts differ only by multiples of 2π. By §41, 3,

(4) $\text{Log}\,(-1) = \pi i,\ \text{Log}\,i = \tfrac{1}{2}\pi i,\ \text{Log}\,(-i) = -\tfrac{1}{2}\pi i.$

The familiar laws for operating with natural logarithms, namely

$$\log\,(z_1 z_2) = \log z_1 + \log z_2\,,$$
$$(z_1 \neq 0,\, z_2 \neq 0),$$
(5) $$\log\frac{z_1}{z_2} = \log z_1 - \log z_2\,,$$
$$\log z^k = k\,\log z, \qquad (z \neq 0,\, k \text{ integral}),$$

are now valid because they follow formally from the given definition of the logarithm. However, because of the ambiguity of the symbol "log", they are to be understood in the sense that each value of one side is also contained among the values of the other side, except in the last of the three laws, where all that we can say in general is that every value on the right is contained among the values on the left.[75]

Since, when considering the function $w = \log z$, we are dealing merely with the inverse of the relation $e^w = z$, the arguments in §41, 8 immediately furnish us with all details of the mapping effected by the logarithmic function. We need only make therein a suitable interchange of z and w. Thus, by means of the principal value $w = \text{Log}\,z$, the interior of the z-plane cut along the negative axis of reals, is mapped in a one-to-one and, without exception, conformal manner on the interior of the strip $-\pi < \Im(w) < +\pi$ in the w-plane.

It can be shown, in the following manner, that the principal value of $\log\,[1/(1 - z)]$, for $|z| < 1$, is represented by the same series as in the real domain; i.e., that

(6) $$\text{Log}\,\frac{1}{1 - z} = \sum_{n=1}^{\infty}\frac{z^n}{n}$$

for $|z| < 1$. Since, for real $|x| < 1$, $\log\,[1/(1 - x)] = \sum_{n=1}^{\infty}(x^n/n)$, the composite function[76]

[75] Cf. p. 29, footnote 24.

[76] We shall find it convenient sometimes to write exp z instead of e^z.

$$\exp\left(\sum_{n=1}^{\infty} \frac{x^n}{n}\right) = \frac{1}{1-x}.$$

This means more precisely: If we substitute (see the last considerations in §30) the power series

$$y = x + \frac{x^2}{2} + \cdots + \frac{x^n}{n} + \cdots$$

in the power series

$$(7) \qquad 1 + y + \frac{y^2}{2!} + \cdots + \frac{y^n}{n!} + \cdots,$$

the result obtained is the geometric series $\sum x^n$, since this is the expansion of $1/(1-x)$. This operation of substituting one power series in another is a purely formal one, so that we must also get the geometric series if we substitute the power series

$$w = z + \frac{z^2}{2} + \cdots + \frac{z^n}{n} + \cdots$$

in the exponential series

$$(8) \qquad 1 + w + \frac{w^2}{2!} + \cdots + \frac{w^n}{n!} + \cdots.$$

Hence, for $|z| < 1$,

$$\exp\left(\sum_{n=1}^{\infty} \frac{z^n}{n}\right) = \frac{1}{1-z}; \qquad \text{i.e.,} \qquad \sum_{n=1}^{\infty} \frac{z^n}{n} = \log \frac{1}{1-z}.$$

Now, that the series represents actually the principal value, that, in other words, $-\pi < \Im\left(\sum_{n=1}^{\infty} (z^n/n)\right) \leqq +\pi$ for $|z| < 1$, can be seen as follows: This imaginary part of the sum of the series is, by (3), equal to one of the values of am $[1/(1-z)]$, and hence, equal to $\psi + 2k\pi$, where ψ denotes the principal value of this amplitude. For $|z| < 1$, this principal vaue ψ satisfies the condition $-\pi/2 < \psi < +\pi/2$. Now, for $z = 0$, the series certainly yields the principal value Log 1; we must therefore take $k = 0$. And since $\Im(\sum(z^n/n))$ varies continuously with z in $|z| < 1$, k must remain equal to 0.

From (6) we obtain, by replacing z by $-z$ and changing sign:

$$(9) \qquad \text{Log}\,(1+z) = z - \frac{z^2}{2} + \frac{z^3}{3} - + \cdots, \qquad |z| < 1.$$

The sum of (6) and (9) gives the series

$$(10) \qquad \text{Log}\,\frac{1+z}{1-z} = 2\left[z + \frac{z^3}{3} + \frac{z^5}{5} + \cdots\right], \qquad |z| < 1.^{77}$$

Finally, we determine the *derivative* of the logarithmic function. Let z lie in the interior of the plane which has been cut as above. Then, for all sufficiently small $h \neq 0$,

$$\text{Log}\,(z+h) - \text{Log}\,z = \text{Log}\left(1 + \frac{h}{z}\right).$$

If we set $h/z = h'$ for brevity, we obtain

$$\frac{\text{Log}\,(z+h) - \text{Log}\,z}{h} = \frac{1}{z} \cdot \frac{\text{Log}\,(1+h')}{h'}.$$

The series (9) shows, however, that, as $h' \to 0$, the last quotient tends to 1. Hence, everywhere in the interior of the cut plane,

$$\frac{d\,\text{Log}\,z}{dz} = \frac{1}{z}, \qquad \text{and, consequently, also} \qquad \frac{d\,\log z}{dz} = \frac{1}{z}.$$

46. The cyclometric functions

On the basis of the results in §§42, 43, we can undertake to find the inverses of the trigonometric functions, just as in the preceding paragraph we considered the inverse of the exponential function. This leads to the *cyclometric* (or *inverse trigonometric*) *functions*. We confine ourselves to the inverses of sin z and tan z.

I. According to §42, 6 and 7, the equation sin $w = z$ has, for arbitrarily given z, an infinite number of solutions w. If w^* is a definite one of these, all remaining ones are contained in the two formulas

$$(1) \quad w^* + 2k\pi \quad \text{and} \quad \pi - w^* + 2k\pi, \quad (k = 0, \pm 1, \pm 2, \ldots).$$

---

[77] To see that this series also represents the principal value on the left, requires a little reflection, which we leave to the reader.

From this we easily infer that there is always *precisely one* of these values which lies in the strip

(2) $$-\frac{\pi}{2} \leqq \Re(w) \leqq +\frac{\pi}{2},$$

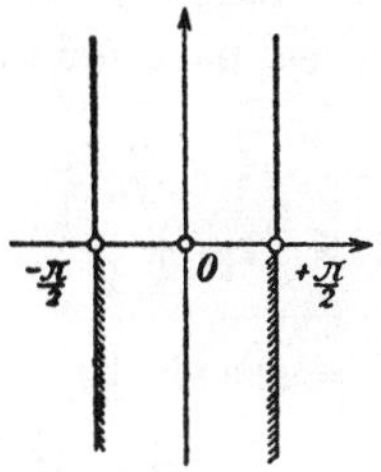

FIGURE 23

provided that we omit from this strip that part of the boundary which lies below the axis of reals (see Fig. 23). In this sense, then, the sine function possesses a single-valued inverse, which is denoted by

(3) $$w = \text{arc sin } z.$$

We call it, more precisely, the *principal value* of the function, terming all other values (1) its subsidiary values. Since (cf. §42, 7) the equation sin $w = z$ is synonymous with the equation $e^{iw} = iz + \sqrt{1 - z^2}$,

(4) $$\text{arc sin } z = \frac{1}{i} \log (iz + \sqrt{1 - z^2}).$$

However, this equation must again be understood to mean that every value of one side is contained among the values of the other side. Thus, whereas sin z can be expressed in terms of the exponential function, the function arc sin z, inversely, can be expressed essentially in terms of the logarithmic function.

Considerations entirely analogous to those in §42 now show, that the series expansion of the real arc-sin-function, known to us from the real domain, must remain valid also in the complex domain. Hence, for $|z| < 1$,

$$\text{arc sin } z = z + \frac{1}{2} \cdot \frac{z^3}{3} + \frac{1 \cdot 3}{2 \cdot 4} \cdot \frac{z^5}{5} + \frac{1 \cdot 3 \cdot 5}{2 \cdot 4 \cdot 6} \cdot \frac{z^7}{7} + \cdots ,$$

(5)

$$(\,|\,z\,|\,<\,1).$$

This series represents actually the principal value of our function. For, by its definition, this is the case if, and only if, the sum of the series (5) has a real part less than $\pi/2$. But we have, in fact,

$$\Re\left(z + \frac{1}{2}\frac{z^3}{3} + \cdots \right) \leqq |z| + \frac{1}{2}\frac{|z|^3}{3} + \cdots$$

$$= \text{arc sin } |z| < \text{arc sin } 1 = \frac{\pi}{2}.$$

We know, moreover, from the real domain, that the representation (5) is still valid for $z = +1$, and hence, that

(6) $$\frac{\pi}{2} = 1 + \frac{1}{2} \cdot \frac{1}{3} + \frac{1 \cdot 3}{2 \cdot 4} \cdot \frac{1}{5} + \cdots .$$

II. According to §43, 4, the equation $\tan w = z$ has, for arbitrarily given $z \neq \pm i$, always one solution w such that $-\pi/2 < \Re(w) \leqq +\pi/2$. We designate this solution as the principal value of the function

(7) $$w = \text{arc tan } z.$$

All remaining solutions of the same equation are obtained from the principal solution by the addition of arbitrary integral multiples of π, and constitute the subsidiary values of this function. Since the equation $\tan w = z$ is synonymous with the equation

$$\frac{1}{i} \frac{e^{iw} - e^{-iw}}{e^{iw} + e^{-iw}} = z \quad \text{or} \quad e^{2iw} = \frac{1 + iz}{1 - iz},$$

we have in

(8) $$w = \text{arc tan } z = \frac{1}{2i} \log \frac{1 + iz}{1 - iz}$$

a representation of the arc-tan function in terms of the logarithm. Equation (8) is, naturally, again to be understood in the sense that every value of one side is contained among the values of the other side.

Finally, the same considerations as in the preceding cases show that, for $|z| < 1$,

$$(9) \qquad \arctan z = z - \frac{z^3}{3} + \frac{z^5}{5} - + \cdots , \qquad (|z| < 1), —$$

an expansion which, in virtue of (8), can also immediately be derived from the expansion §45 (10). This derivation shows also that the series in (9) again represents actually the principal value of arc tan z. For, this series is obtained if we substitute the series (6) and (9) of §45 for the respective logarithms on the right-hand side in

$$w = \frac{1}{2i} \log (1 + iz) + \frac{1}{2i} \log \frac{1}{1 - iz}.$$

Since these last two series have a sum whose imaginary part lies between $-\pi/2$ and $+\pi/2$, the series in (9) has a real part which likewise lies between $-\pi/2$ and $+\pi/2$.

47. *The binomial series and the general power*

In the real domain, by the *binomial series* we mean the series

$$(1) \qquad \sum_{n=0}^{\infty} \binom{\alpha}{n} x^n = 1 + \alpha x + \frac{\alpha(\alpha - 1)}{1 \cdot 2} x^2 + \cdots$$
$$+ \frac{\alpha(\alpha - 1) \cdots (\alpha - n + 1)}{1 \cdot 2 \cdots n} x^n + \cdots .$$

It is convergent for $|x| < 1$, as is shown, without difficulty, by the ratio test; α here may denote an arbitrary real number. The sum of this series is the power

$$(2) \qquad (1 + x)^\alpha$$

of the (for $|x| < 1$) *positive* base $(1 + x)$ to the exponent α; this power, in the real domain, is fully uniquely defined (as a positive value). We shall see that these facts, too, in the main, continue to hold if we allow x and α to be complex numbers. To this end we require, first, the following

DEFINITION. *Let b be a non-zero complex number, and let a be a completely arbitrary complex number. Then, by the (general) power b^a we mean every value given by the formula*

(3) $$b^a = e^{a \log b}.$$

That one of the values (3) which is obtained by taking for log b, there, its principal value, is designated as the principal value of b^a.

Thus, e.g.,

$$i^i = e^{i \log i} = e^{i[(\pi i/2) + 2k\pi i]} = e^{-(\pi/2) - 2k\pi}, \quad (k = 0, \pm 1, \cdots).$$

Of these infinitely many values of i^i (all of which are real!), $e^{-\pi/2}$ is the principal value.[78]

Now, in the real domain, for $|x| < 1$ and real α we have, on the one hand,

$$(1 + x)^\alpha = e^{\alpha \log(1+x)} = \exp\left\{\alpha\left(x - \frac{x^2}{2} + \frac{x^3}{3} - + \cdots\right)\right\},$$

and, on the other hand,

$$(1 + x)^\alpha = \sum_{n=0}^{\infty} \binom{\alpha}{n} x^n.$$

This means: If we substitute the series $y = \alpha\left(x - \dfrac{x^2}{2} + \cdots\right)$

in the exponential series for e^y, and arrange in powers of x in accordance with §30, we obtain the binomial series. This purely formal operation remains valid, naturally, if α and z denote complex numbers. The power series

(4) $$w = \alpha\left(z - \frac{z^2}{2} + \frac{z^3}{3} - + \cdots\right), \quad (\alpha \text{ arbitrary, complex}),$$

---

[78]Without proof, we call attention to the fact that, for the general power defined above, the old rules of operation

$$b^a \cdot b^{a'} = b^{a+a'} \quad \text{and} \quad (b^a)^{a'} = b^{aa'}$$

do *not* hold any more,—not even in the extended sense that every value of one side is contained among the values of the other side. On the contrary, in each of these two equations, the left-hand side represents *more* values than the right-hand side.

is absolutely convergent for $|z| < 1$ and is then the principal value of $\alpha \log (1 + z)$. Hence, if we substitute this series in the exponential series for e^w, and arrange in powers of z, we must again obtain the *binomial series*

(5)
$$\sum_{n=0}^{\infty} \binom{\alpha}{n} z^n = 1 + \alpha z + \frac{\alpha(\alpha - 1)}{1 \cdot 2} z^2 + \cdots$$

$$+ \frac{\alpha(\alpha - 1) \cdots (\alpha - n + 1)}{1 \cdot 2 \cdots n} z^n + \cdots ,$$

and its sum must be the principal value of

(6)
$$(1 + z)^{\alpha}, —$$

provided that the effected rearrangement in the sense of §30 is permissible. But this is certainly the case, because the exponential series converges everywhere, and the series (4) remains convergent for $|z| < 1$ if all its terms are replaced by their respective absolute values.[79] Thus, for all $|z| < 1$ and arbitrary, complex α, the series (5) represents the power (6),—the latter is developable, as a function of z, in the power series (5).

[79]From this it follows automatically, that the binomial series is absolutely convergent for $|z| < 1$. This is also very easy to show directly, with the aid of the ratio test, since, as $n \to \infty$,

$$\left| \binom{\alpha}{n+1} z^{n+1} \middle/ \binom{\alpha}{n} z^n \right| = \left| \frac{\alpha - n}{n + 1} \cdot z \right| \to |z|,$$

and it is assumed that $|z| < 1$.

BIBLIOGRAPHY

T. J. I'ᴀ. Bromwich, *An Introduction to the Theory of Infinite Series*, 2d edition, London, 1947.

H. Burkhardt, *Funktionentheoretische Vorlesungen*,
vol. I, 1: *Algebraische Analysis*,
vol. I, 2: *Einführung in die Theorie der analytischen Funktionen einer komplexen Veränderlichen*,
edited by G. Faber, Berlin, 1920–21.

———, *Theory of Functions of a Complex Variable*, translated by S. E. Rasor from the 4th German edition, Boston, 1913.

H. Falckenberg, *Elementare Reihenlehre*, 2d edition, Berlin (Sammlung Göschen no. 943), 1944.

———, *Komplexe Reihen nebst Aufgaben über reelle und komplexe Reihen*, new impression, revised, Berlin (Sammlung Göschen no. 1027), 1944.

G. H. Hardy, *A Course of Pure Mathematics*, 9th edition, New York, 1947.

K. Knopp, *Theorie und Anwendung der unendlichen Reihen*, 4th edition, Berlin and Heidelberg, 1947.

———, *Theory and Application of Infinite Series*, translated by R. C. Young from the 2d German edition, London and Glasgow, 1928.

H. v. Mangoldt and K. Knopp, *Einführung in die höhere Mathematik*, 9th edition, vols. I and II, Zürich, 1948.

N. Nielsen, *Elemente der Funktionentheorie*, Leipzig and Berlin, 1911.

O. Perron, *Irrationalzahlen*, 2d edition, New York (reprint), 1948.

A. Pringsheim and G. Faber, *Algebraische Analysis*, Enzyklopädie der mathematischen Wissenschaften, vol. II, C, 1, Leipzig, 1909.

136

INDEX

CATALOGUE OF DOVER BOOKS

MATHEMATICS—INTERMEDIATE TO ADVANCED

General

INTRODUCTION TO APPLIED MATHEMATICS, Francis D. Murnaghan. A practical and thoroughly sound introduction to a number of advanced branches of higher mathematics. Among the selected topics covered in detail are: vector and matrix analysis, partial and differential equations, integral equations, calculus of variations, Laplace transform theory, the vector triple product, linear vector functions, quadratic and bilinear forms, Fourier series, spherical harmonics, Bessel functions, the Heaviside expansion formula, and many others. Extremely useful book for graduate students in physics, engineering, chemistry, and mathematics. Index. 111 study exercises with answers. 41 illustrations. ix + 389pp. 5⅜ x 8½.
S1042 Paperbound **$2.00**

OPERATIONAL METHODS IN APPLIED MATHEMATICS, H. S. Carslaw and J. C. Jaeger. Explanation of the application of the Laplace Transformation to differential equations, a simple and effective substitute for more difficult and obscure operational methods. Of great practical value to engineers and to all workers in applied mathematics. Chapters on: Ordinary Linear Differential Equations with Constant Coefficients;; Electric Circuit Theory; Dynamical Applications; The Inversion Theorem for the Laplace Transformation; Conduction of Heat; Vibrations of Continuous Mechanical Systems; Hydrodynamics; Impulsive Functions; Chains of Differential Equations; and other related matters. 3 appendices. 153 problems, many with answers. 22 figures. xvi + 359pp. 5⅜ x 8½.
S1011 Paperbound **$2.25**

APPLIED MATHEMATICS FOR RADIO AND COMMUNICATIONS ENGINEERS, C. E. Smith. No extraneous material here!—only the theories, equations, and operations essential and immediately useful for radio work. Can be used as refresher, as handbook of applications and tables, or as full home-study course. Ranges from simplest arithmetic through calculus, series, and wave forms, hyperbolic trigonometry, simultaneous equations in mesh circuits, etc. Supplies applications right along with each math topic discussed. 22 useful tables of functions, formulas, logs, etc. Index. 166 exercises, 140 examples, all with answers. 95 diagrams. Bibliography. x + 336pp. 5⅜ x 8.
S141 Paperbound **$1.75**

Algebra, group theory, determinants, sets, matrix theory

ALGEBRAS AND THEIR ARITHMETICS, L. E. Dickson. Provides the foundation and background necessary to any advanced undergraduate or graduate student studying abstract algebra. Begins with elementary introduction to linear transformations, matrices, field of complex numbers; proceeds to order, basal units, modulus, quaternions, etc.; develops calculus of linears sets, describes various examples of algebras including invariant, difference, nilpotent, semi-simple. "Makes the reader marvel at his genius for clear and profound analysis," Amer. Mathematical Monthly. Index. xii + 241pp. 5⅜ x 8.
S616 Paperbound **$1.50**

THE THEORY OF EQUATIONS WITH AN INTRODUCTION TO THE THEORY OF BINARY ALGEBRAIC FORMS, W. S. Burnside and A. W. Panton. Extremely thorough and concrete discussion of the theory of equations, with extensive detailed treatment of many topics curtailed in later texts. Covers theory of algebraic equations, properties of polynomials, symmetric functions, derived functions, Horner's process, complex numbers and the complex variable, determinants and methods of elimination, invariant theory (nearly 100 pages), transformations, introduction to Galois theory, Abelian equations, and much more. Invaluable supplementary work for modern students and teachers. 759 examples and exercises. Index in each volume. Two volume set. Total of xxiv + 604pp. 5⅜ x 8.
S714 Vol I Paperbound **$1.85**
S715 Vol II Paperbound **$1.85**
The set **$3.70**

COMPUTATIONAL METHODS OF LINEAR ALGEBRA, V. N. Faddeeva, translated by **C. D. Benster.** First English translation of a unique and valuable work, the only work in English presenting a systematic exposition of the most important methods of linear algebra—classical and contemporary. Shows in detail how to derive numerical solutions of problems in mathematical physics which are frequently connected with those of linear algebra. Theory as well as individual practice. Part I surveys the mathematical background that is indispensable to what follows. Parts II and III, the conclusion, set forth the most important methods of solution, for both exact and iterative groups. One of the most outstanding and valuable features of this work is the 23 tables, double and triple checked for accuracy. These tables will not be found elsewhere. Author's preface. Translator's note. New bibliography and index. x + 252pp. 5⅜ x 8.
S424 Paperbound **$1.95**

ALGEBRAIC EQUATIONS, E. Dehn. Careful and complete presentation of Galois' theory of algebraic equations; theories of Lagrange and Galois developed in logical rather than historical form, with a more thorough exposition than in most modern books. Many concrete applications and fully-worked-out examples. Discusses basic theory (very clear exposition of the symmetric group); isomorphic, transitive, and Abelian groups; applications of Lagrange's and Galois' theories; and much more. Newly revised by the author. Index. List of Theorems. xi + 208pp. 5⅜ x 8.
S697 Paperbound **$1.45**

THEORY AND APPLICATIONS OF FINITE GROUPS, G. A. Miller, H. F. Blichfeldt, L. E. Dickson. Unusually accurate and authoritative work, each section prepared by a leading specialist: Miller on substitution and abstract groups, Blichfeldt on finite groups of linear homogeneous transformations, Dickson on applications of finite groups. Unlike more modern works, this gives the concrete basis from which abstract group theory arose. Includes Abelian groups, prime-power groups, isomorphisms, matrix forms of linear transformations, Sylow groups, Galois' theory of algebraic equations, duplication of a cube, trisection of an angle, etc. 2 Indexes. 267 problems. xvii + 390pp. 5⅜ x 8. S216 Paperbound **$2.00**

THE THEORY OF DETERMINANTS, MATRICES, AND INVARIANTS, H. W. Turnbull. Important study includes all salient features and major theories. 7 chapters on determinants and matrices cover fundamental properties, Laplace identities, multiplication, linear equations, rank and differentiation, etc. Sections on invariants gives general properties, symbolic and direct methods of reduction, binary and polar forms, general linear transformation, first fundamental theorem, multilinear forms. Following chapters study development and proof of Hilbert's Basis Theorem, Gordan-Hilbert Finiteness Theorem, Clebsch's Theorem, and include discussions of apolarity, canonical forms, geometrical interpretations of algebraic forms, complete system of the general quadric, etc. New preface and appendix. Bibliography. xviii + 374pp. 5⅜ x 8. S699 Paperbound **$2.25**

AN INTRODUCTION TO THE THEORY OF CANONICAL MATRICES, H. W. Turnbull and A. C. Aitken. All principal aspects of the theory of canonical matrices, from definitions and fundamental properties of matrices to the practical applications of their reduction to canonical form. Beginning with matrix multiplications, reciprocals, and partitioned matrices, the authors go on to elementary transformations and bilinear and quadratic forms. Also covers such topics as a rational canonical form for the collineatory group, congruent and conjunctive transformation for quadratic and hermitian forms, unitary and orthogonal transformations, canonical reduction of pencils of matrices, etc. Index. Appendix. Historical notes at chapter ends. Bibliographies. 275 problems. xiv + 200pp. 5⅜ x 8. S177 Paperbound **$1.55**

A TREATISE ON THE THEORY OF DETERMINANTS, T. Muir. Unequalled as an exhaustive compilation of nearly all the known facts about determinants up to the early 1930's. Covers notation and general properties, row and column transformation, symmetry, compound determinants, adjugates, rectangular arrays and matrices, linear dependence, gradients, Jacobians, Hessians, Wronskians, and much more. Invaluable for libraries of industrial and research organizations as well as for student, teacher, and mathematician; very useful in the field of computing machines. Revised and enlarged by W. H. Metzler. Index. 485 problems and scores of numerical examples. iv + 766pp. 5⅜ x 8. S670 Paperbound **$3.00**

THEORY OF DETERMINANTS IN THE HISTORICAL ORDER OF DEVELOPMENT, Sir Thomas Muir. Unabridged reprinting of this complete study of 1,859 papers on determinant theory written between 1693 and 1900. Most important and original sections reproduced, valuable commentary on each. No other work is necessary for determinant research: all types are covered— each subdivision of the theory treated separately; all papers dealing with each type are covered; you are told exactly what each paper is about and how important its contribution is. Each result, theory, extension, or modification is assigned its own identifying numeral so that the full history may be more easily followed. Includes papers on determinants in general, determinants and linear equations, symmetric determinants, alternants, recurrents, determinants having invariant factors, and all other major types. "A model of what such histories ought to be," NATURE. "Mathematicians must ever be grateful to Sir Thomas for his monumental work," AMERICAN MATH MONTHLY. Four volumes bound as two. Indices. Bibliographies. Total of lxxxiv + 1977pp. 5⅜ x 8. S672-3 The set, Clothbound **$12.50**

Calculus and function theory, Fourier theory, infinite series, calculus of variations, real and complex functions

FIVE VOLUME "THEORY OF FUNCTIONS' SET BY KONRAD KNOPP

This five-volume set, prepared by Konrad Knopp, provides a complete and readily followed account of theory of functions. Proofs are given concisely, yet without sacrifice of completeness or rigor. These volumes are used as texts by such universities as M.I.T., University of Chicago, N. Y. City College, and many others. "Excellent introduction . . . remarkably readable, concise, clear, rigorous," JOURNAL OF THE AMERICAN STATISTICAL ASSOCIATION.

ELEMENTS OF THE THEORY OF FUNCTIONS, Konrad Knopp. This book provides the student with background for further volumes in this set, or texts on a similar level. Partial contents: foundations, system of complex numbers and the Gaussian plane of numbers, Riemann sphere of numbers, mapping by linear functions, normal forms, the logarithm, the cyclometric functions and binomial series. "Not only for the young student, but also for the student who knows all about what is in it," MATHEMATICAL JOURNAL. Bibliography. Index. 140pp. 5⅜ x 8. S154 Paperbound **$1.35**

THEORY OF FUNCTIONS, PART I, Konrad Knopp. With volume II, this book provides coverage of basic concepts and theorems. Partial contents: numbers and points, functions of a complex variable, integral of a continuous function, Cauchy's integral theorem, Cauchy's integral formulae, series with variable terms, expansion of analytic functions in power series, analytic continuation and complete definition of analytic functions, entire transcendental functions, Laurent expansion, types of singularities. Bibliography. Index. vii + 146pp. 5⅜ x 8. S156 Paperbound **$1.35**

THEORY OF FUNCTIONS, PART II, Konrad Knopp. Application and further development of general theory, special topics. Single valued functions, entire, Weierstrass, Meromorphic functions. Riemann surfaces. Algebraic functions. Analytical configuration, Riemann surface. Bibliography. Index. x + 150pp. 5⅜ x 8.   S157 Paperbound **$1.35**

PROBLEM BOOK IN THE THEORY OF FUNCTIONS, VOLUME 1, Konrad Knopp. Problems in elementary theory, for use with Knopp's THEORY OF FUNCTIONS, or any other text, arranged according to increasing difficulty. Fundamental concepts, sequences of numbers and infinite series, complex variable, integral theorems, development in series, conformal mapping. 182 problems. Answers. viii + 126pp. 5⅜ x 8.   S158 Paperbound **$1.35**

PROBLEM BOOK IN THE THEORY OF FUNCTIONS, VOLUME 2, Konrad Knopp. Advanced theory of functions, to be used either with Knopp's THEORY OF FUNCTIONS, or any other comparable text. Singularities, entire & meromorphic functions, periodic, analytic, continuation, multiple-valued functions, Riemann surfaces, conformal mapping. Includes a section of additional elementary problems. "The difficult task of selecting from the immense material of the modern theory of functions the problems just within the reach of the beginner is here masterfully accomplished," AM. MATH. SOC. Answers. 138pp. 5⅜ x 8.   S159 Paperbound **$1.35**

A COURSE IN MATHEMATICAL ANALYSIS, Edouard Goursat. Trans. by E. R. Hedrick, O. Dunkel. Classic study of fundamental material thoroughly treated. Exceptionally lucid exposition of wide range of subject matter for student with 1 year of calculus. Vol. 1: Derivatives and Differentials, Definite Integrals, Expansion in Series, Applications to Geometry. Problems. Index. 52 illus. 556pp. Vol. 2, Part I: Functions of a Complex Variable, Conformal Representations, Doubly Periodic Functions, Natural Boundaries, etc. Problems. Index. 38 illus. 269pp. Vol. 2, Part 2: Differential Equations, Cauchy-Lipschitz Method, Non-linear Differential Equations, Simultaneous Equations, etc. Problems. Index. 308pp. 5⅜ x 8.

Vol. 1 S554 Paperbound **$2.50**
Vol. 2 part 1 S555 Paperbound **$1.85**
Vol. 2 part 2 S556 Paperbound **$1.85**
3 vol. set **$6.20**

MODERN THEORIES OF INTEGRATION, H. Kestelman. Connected and concrete coverage, with fully-worked-out proofs for every step. Ranges from elementary definitions through theory of aggregates, sets of points, Riemann and Lebesgue integration, and much more. This new revised and enlarged edition contains a new chapter on Riemann-Stieltjes integration, as well as a supplementary section of 186 exercises. Ideal for the mathematician, student, teacher, or self-studier. Index of Definitions and Symbols. General Index. Bibliography. x + 310pp. 5⅝ x 8⅜.   S572 Paperbound **$2.25**

THEORY OF MAXIMA AND MINIMA, H. Hancock. Fullest treatment ever written; only work in English with extended discussion of maxima and minima for functions of 1, 2, or n variables, problems with subsidiary constraints, and relevant quadratic forms. Detailed proof of each important theorem. Covers the Scheeffer and von Dantscher theories, homogeneous quadratic forms, reversion of series, fallacious establishment of maxima and minima, etc. Unsurpassed treatise for advanced students of calculus, mathematicians, economists, statisticians. Index. 24 diagrams. 39 problems, many examples. 193pp. 5⅜ x 8.   S665 Paperbound **$1.50**

AN ELEMENTARY TREATISE ON ELLIPTIC FUNCTIONS, A. Cayley. Still the fullest and clearest text on the theories of Jacobi and Legendre for the advanced student (and an excellent supplement for the beginner). A masterpiece of exposition by the great 19th century British mathematician (creator of the theory of matrices and abstract geometry), it covers the addition-theory, Landen's theorem, the 3 kinds of elliptic integrals, transformations, the q-functions, reduction of a differential expression, and much more. Index. xii + 386pp. 5⅜ x 8.
                                                                                S728 Paperbound **$2.00**

THE APPLICATIONS OF ELLIPTIC FUNCTIONS, A. G. Greenhill. Modern books forego detail for sake of brevity—this book offers complete exposition necessary for proper understanding, use of elliptic integrals. Formulas developed from definite physical, geometric problems; examples representative enough to offer basic information in widely useable form. Elliptic integrals, addition theorem, algebraical form of addition theorem, elliptic integrals of 2nd, 3rd kind, double periodicity, resolution into factors, series, transformation, etc. Introduction. Index. 25 illus. xi + 357pp. 5⅜ x 8.   S603 Paperbound **$1.75**

THE THEORY OF FUNCTIONS OF REAL VARIABLES, James Pierpont. A 2-volume authoritative exposition, by one of the foremost mathematicians of his time. Each theorem stated with all conditions, then followed by proof. No need to go through complicated reasoning to discover conditions added without specific mention. Includes a particularly complete, rigorous presentation of theory of measure; and Pierpont's own work on a theory of Lebesgue integrals, and treatment of area of a curved surface. Partial contents, Vol. 1: rational numbers, exponentials, logarithms, point aggregates, maxima, minima, proper integrals, improper integrals, multiple proper integrals, continuity, discontinuity, indeterminate forms. Vol. 2: point sets, proper integrals, series, power series, aggregates, ordinal numbers, discontinuous functions, sub-, infra-uniform convergence, much more. Index. 95 illustrations. 1229pp. 5⅜ x 8.   S558-9, 2 volume set, paperbound **$5.20**

FUNCTIONS OF A COMPLEX VARIABLE, James Pierpont. Long one of best in the field. A thorough treatment of fundamental elements, concepts, theorems. A complete study, rigorous, detailed, with carefully selected problems worked out to illustrate each topic. Partial contents: arithmetical operations, real term series, positive term series, exponential functions, integration, analytic functions, asymptotic expansions, functions of Weierstrass, Legendre, etc. Index. List of symbols. 122 illus. 597pp. 5⅜ x 8. S560 Paperbound **$2.45**

MODERN OPERATIONAL CALCULUS: WITH APPLICATIONS IN TECHNICAL MATHEMATICS, N. W. McLachlan. An introduction to modern operational calculus based upon the Laplace transform, applying it to the solution of ordinary and partial differential equations. For physicists, engineers, and applied mathematicians. Partial contents: Laplace transform, theorems or rules of the operational calculus, solution of ordinary and partial linear differential equations with constant coefficients, evaluation of integrals and establishment of mathematical relationships, derivation of Laplace transforms of various functions, etc. Six appendices deal with Heaviside's unit function, etc. Revised edition. Index. Bibliography. xiv + 218pp. 5⅜ x 8½. S192 Paperbound **$1.75**

ADVANCED CALCULUS, E. B. Wilson. An unabridged reprinting of the work which continues to be recognized as one of the most comprehensive and useful texts in the field. It contains an immense amount of well-presented, fundamental material, including chapters on vector functions, ordinary differential equations, special functions, calculus of variations, etc., which are excellent introductions to these areas. For students with only one year of calculus, more than 1300 exercises cover both pure math and applications to engineering and physical problems. For engineers, physicists, etc., this work, with its 54 page introductory review, is the ideal reference and refresher. Index. ix + 566pp. 5⅜ x 8. S504 Paperbound **$2.45**

ASYMPTOTIC EXPANSIONS, A. Erdélyi. The only modern work available in English, this is an unabridged reproduction of a monograph prepared for the Office of Naval Research. It discusses various procedures for asymptotic evaluation of integrals containing a large parameter and solutions of ordinary linear differential equations. Bibliography of 71 items. vi + 108pp. 5⅜ x 8. S318 Paperbound **$1.35**

INTRODUCTION TO ELLIPTIC FUNCTIONS: with applications, F. Bowman. Concise, practical introduction to elliptic integrals and functions. Beginning with the familiar trigonometric functions, it requires nothing more from the reader than a knowledge of basic principles of differentiation and integration. Discussion confined to the Jacobian functions. Enlarged bibliography. Index. 173 problems and examples. 56 figures, 4 tables. 115pp. 5⅜ x 8. S922 Paperbound **$1.25**

ON RIEMANN'S THEORY OF ALGEBRAIC FUNCTIONS AND THEIR INTEGRALS: A SUPPLEMENT TO THE USUAL TREATISES, Felix Klein. Klein demonstrates how the mathematical ideas in Riemann's work on Abelian integrals can be arrived at by thinking in terms of the flow of electric current on surfaces. Intuitive explanations, not detailed proofs given in an extremely clear exposition, concentrating on the kinds of functions which can be defined on Riemann surfaces. Also useful as an introduction to the origins of topological problems. Complete and unabridged. Approved translation by Frances Hardcastle. New introduction. 43 figures. Glossary. xii + 76pp. 5⅜ x 8½. S1072 Paperbound **$1.25**

COLLECTED WORKS OF BERNHARD RIEMANN. This important source book is the first to contain the complete text of both 1892 Werke and the 1902 supplement, unabridged. It contains 31 monographs, 3 complete lecture courses, 15 miscellaneous papers, which have been of enormous importance in relativity, topology, theory of complex variables, and other areas of mathematics. Edited by R. Dedekind, H. Weber, M. Noether, W. Wirtinger. German text. English introduction by Hans Lewy. 690pp. 5⅜ x 8. S226 Paperbound **$3.75**

THE TAYLOR SERIES, AN INTRODUCTION TO THE THEORY OF FUNCTIONS OF A COMPLEX VARIABLE, P. Dienes. This book investigates the entire realm of analytic functions. Only ordinary calculus is needed, except in the last two chapters. Starting with an introduction to real variables and complex algebra, the properties of infinite series, elementary functions, complex differentiation and integration are carefully derived. Also biuniform mapping, a thorough two part discussion of representation and singularities of analytic functions, overconvergence and gap theorems, divergent series, Taylor series on its circle of convergence, divergence and singularities, etc. Unabridged, corrected reissue of first edition. Preface and index. 186 examples, many fully worked out. 67 figures. xii + 555pp. 5⅜ x 8. S391 Paperbound **$2.75**

INTRODUCTION TO BESSEL FUNCTIONS, Frank Bowman. A rigorous self-contained exposition providing all necessary material during the development, which requires only some knowledge of calculus and acquaintance with differential equations. A balanced presentation including applications and practical use. Discusses Bessel Functions of Zero Order, of Any Real Order; Modified Bessel Functions of Zero Order; Definite Integrals; Asymptotic Expansions; Bessel's Solution to Kepler's Problem; Circular Membranes; much more. "Clear and straightforward . . . useful not only to students of physics and engineering, but to mathematical students in general," Nature. 226 problems. Short tables of Bessel functions. 27 figures. Index. x + 135pp. 5⅜ x 8. S462 Paperbound **$1.35**

ELEMENTS OF THE THEORY OF REAL FUNCTIONS, J. E. Littlewood. Based on lectures given at Trinity College, Cambridge, this book has proved to be extremely successful in introducing graduate students to the modern theory of functions. It offers a full and concise coverage of classes and cardinal numbers, well-ordered series, other types of series, and elements of the theory of sets of points. 3rd revised edition. vii + 71pp. 5⅜ x 8.

S171 Clothbound **$2.85**
S172 Paperbound **$1.25**

TRANSCENDENTAL AND ALGEBRAIC NUMBERS, A. O. Gelfond. First English translation of work by leading Soviet mathematician. Thue-Siegel theorem, its p-adic analogue, on approximation of algebraic numbers by numbers in fixed algebraic field; Hermite-Lindemann theorem on transcendency of Bessel functions, solutions of other differential equations; Gelfond-Schneider theorem on transcendency of alpha to power beta; Schneider's work on elliptic functions, with method developed by Gelfond. Translated by L. F. Boron. Index. Bibliography. 200pp. 5⅜ x 8.

S615 Paperbound **$1.75**

ELLIPTIC INTEGRALS, H. Hancock. Invaluable in work involving differential equations containing cubics or quartics under the root sign, where elementary calculus methods are inadequate. Practical solutions to problems that occur in mathematics, engineering, physics: differential equations requiring integration of Lamé's, Briot's, or Bouquet's equations; determination of arc of ellipse, hyperbola, lemniscate; solutions of problems in elastica; motion of a projectile under resistance varying as the cube of the velocity; pendulums; many others. Exposition is in accordance with Legendre-Jacobi theory and includes rigorous discussion of Legendre transformations. 20 figures. 5 place table. Index. 104pp. 5⅛ x 8.

S484 Paperbound **$1.25**

LECTURES ON THE THEORY OF ELLIPTIC FUNCTIONS, H. Hancock. Reissue of the only book in English with so extensive a coverage, especially of Abel, Jacobi, Legendre, Weierstrasse, Hermite, Liouville, and Riemann. Unusual fullness of treatment, plus applications as well as theory, in discussing elliptic function (the universe of elliptic integrals originating in works of Abel and Jacobi), their existence, and ultimate meaning. Use is made of Riemann to provide the most general theory. 40 page table of formulas. 76 figures. xxiii + 498pp.

S483 Paperbound **$2.55**

THE THEORY AND FUNCTIONS OF A REAL VARIABLE AND THE THEORY OF FOURIER'S SERIES, E. W. Hobson. One of the best introductions to set theory and various aspects of functions and Fourier's series. Requires only a good background in calculus. Provides an exhaustive coverage of: metric and descriptive properties of sets of points; transfinite numbers and order types; functions of a real variable; the Riemann and Lebesgue integrals; sequences and series of numbers; power-series; functions representable by series sequences of continuous functions; trigonometrical series; representation of functions by Fourier's series; complete exposition (200pp.) on set theory; and much more. "The best possible guide," Nature. Vol. I: 88 detailed examples, 10 figures. Index. xv + 736pp. Vol. II: 117 detailed examples, 13 figures. Index. x + 780pp. 6⅛ x 9¼.

Vol. I: S387 Paperbound **$3.00**
Vol. II: S388 Paperbound **$3.00**

ALMOST PERIODIC FUNCTIONS, A. S. Besicovitch. This unique and important summary by a well-known mathematician covers in detail the two stages of development in Bohr's theory of almost periodic functions: (1) as a generalization of pure periodicity, with results and proofs; (2) the work done by Stepanoff, Wiener, Weyl, and Bohr in generalizing the theory. Bibliography. xi + 180pp. 5⅜ x 8.

S18 Paperbound **$1.75**

THE ANALYTICAL THEORY OF HEAT, Joseph Fourier. This book, which revolutionized mathematical physics, is listed in the Great Books program, and many other listings of great books. It has been used with profit by generations of mathematicians and physicists who are interested in either heat or in the application of the Fourier integral. Covers cause and reflection of rays of heat, radiant heating, heating of closed spaces, use of trigonometric series in the theory of heat, Fourier integral, etc. Translated by Alexander Freeman. 20 figures. xxii + 466pp. 5⅜ x 8.

S93 Paperbound **$2.50**

AN INTRODUCTION TO FOURIER METHODS AND THE LAPLACE TRANSFORMATION, Philip Franklin. Concentrates upon essentials, enabling the reader with only a working knowledge of calculus to gain an understanding of Fourier methods in a broad sense, suitable for most applications. This work covers complex qualities with methods of computing elementary functions for complex values of the argument and finding approximations by the use of charts; Fourier series and integrals with half-range and complex Fourier series; harmonic analysis; Fourier and Laplace transformations, etc.; partial differential equations with applications to transmission of electricity; etc. The methods developed are related to physical problems of heat flow, vibrations, electrical transmission, electromagnetic radiation, etc. 828 problems with answers. Formerly entitled "Fourier Methods." Bibliography. Index. x + 289pp. 5⅜ x 8.

S452 Paperbound **$2.00**

THE FOURIER INTEGRAL AND CERTAIN OF ITS APPLICATIONS, Norbert Wiener. The only book-length study of the Fourier integral as link between pure and applied math. An expansion of lectures given at Cambridge. Partial contents: Plancherel's theorem, general Tauberian theorem, special Tauberian theorems, generalized harmonic analysis. Bibliography. viii + 201pp. 5⅜ x 8.

S272 Paperbound **$1.50**

INTRODUCTION TO THE THEORY OF FOURIER'S SERIES AND INTEGRALS, H. S. Carslaw. 3rd revised edition. This excellent introduction is an outgrowth of the author's courses at Cambridge. Historical introduction, rational and irrational numbers, infinite sequences and series, functions of a single variable, definite integral, Fourier series, Fourier integrals, and similar topics. Appendixes discuss practical harmonic analysis, periodogram analysis. Lebesgue's theory. Indexes. 84 examples, bibliography. xii + 368pp. 5⅜ x 8.   S48 Paperbound **$2.00**

FOURIER'S SERIES AND SPHERICAL HARMONICS, W. E. Byerly. Continues to be recognized as one of most practical, useful expositions. Functions, series, and their differential equations are concretely explained in great detail; theory is applied constantly to practical problems, which are fully and lucidly worked out. Appendix includes 6 tables of surface zonal harmonics, hyperbolic functions, Bessel's functions. Bibliography. 190 problems, approximately half with answers. ix + 287pp. 5⅜ x 8.   S536 Paperbound **$1.75**

INFINITE SEQUENCES AND SERIES, Konrad Knopp. First publication in any language! Excellent introduction to 2 topics of modern mathematics, designed to give the student background to penetrate farther by himself. Sequences & sets, real & complex numbers, etc. Functions of a real & complex variable. Sequences & series. Infinite series. Convergent power series. Expansion of elementary functions. Numerical evaluation of series. Bibliography. v + 186pp. 5⅜ x 8.   S153 Paperbound **$1.75**

TRIGONOMETRICAL SERIES, Antoni Zygmund. Unique in any language on modern advanced level. Contains carefully organized analyses of trigonometric, orthogonal, Fourier systems of functions, with clear adequate descriptions of summability of Fourier series, proximation theory, conjugate series, convergence, divergence of Fourier series. Especially valuable for Russian, Eastern European coverage. Bibliography. 329pp. 5⅜ x 8.   S290 Paperbound **$1.50**

DICTIONARY OF CONFORMAL REPRESENTATIONS, H. Kober. Laplace's equation in 2 dimensions solved in this unique book developed by the British Admiralty. Scores of geometrical forms & their transformations for electrical engineers, Joukowski aerofoil for aerodynamists. Schwarz-Christoffel transformations for hydrodynamics, transcendental functions. Contents classified according to analytical functions describing transformation. Twin diagrams show curves of most transformations with corresponding regions. Glossary. Topological index. 447 diagrams. 244pp. 6⅛ x 9¼.   S160 Paperbound **$2.00**

CALCULUS OF VARIATIONS, A. R. Forsyth. Methods, solutions, rather than determination of weakest valid hypotheses. Over 150 examples completely worked-out show use of Euler, Legendre, Jacobi, Weierstrass tests for maxima, minima. Integrals with one original dependent variable; with derivatives of 2nd order; two dependent variables, one independent variable; double integrals involving 1 dependent variable, 2 first derivatives; double integrals involving partial derivatives of 2nd order; triple integrals; much more. 50 diagrams. 678pp. 5⅝ x 8⅜.   S622 Paperbound **$2.95**

LECTURES ON THE CALCULUS OF VARIATIONS, O. Bolza. Analyzes in detail the fundamental concepts of the calculus of variations, as developed from Euler to Hilbert, with sharp formulations of the problems and rigorous demonstrations of their solutions. More than a score of solved examples; systematic references for each theorem. Covers the necessary and sufficient conditions; the contributions made by Euler, Du Bois-Reymond, Hilbert, Weierstrass, Legendre, Jacobi, Erdmann, Kneser, and Gauss; and much more. Index. Bibliography. xi + 271pp. 5⅜ x 8.   S218 Paperbound **$1.65**

A TREATISE ON THE CALCULUS OF FINITE DIFFERENCES, G. Boole. A classic in the literature of the calculus. Thorough, clear discussion of basic principles, theorems, methods. Covers MacLaurin's and Herschel's theorems, mechanical quadrature, factorials, periodical constants, Bernoulli's numbers, difference-equations (linear, mixed, and partial), etc. Stresses analogies with differential calculus. 236 problems, answers to the numerical ones. viii + 336pp. 5⅜ x 8.   S695 Paperbound **$1.85**

Prices subject to change without notice.

Dover publishes books on art, music, philosophy, literature, languages, history, social sciences, psychology, handcrafts, orientalia, puzzles and entertainments, chess, pets and gardens, books explaining science, intermediate and higher mathematics, mathematical physics, engineering, biological sciences, earth sciences, classics of science, etc. Write to:

Dept. catrr.
Dover Publications, Inc.
180 Varick Street, N.Y. 14, N.Y.